George Henry Lewes

Studies in animal life

George Henry Lewes

Studies in animal life

ISBN/EAN: 9783337815080

Printed in Europe, USA, Canada, Australia, Japan

Cover: Foto ©ninafisch / pixelio.de

More available books at **www.hansebooks.com**

STUDIES

IN

ANIMAL LIFE.

BY

GEORGE HENRY LEWES,

AUTHOR OF "LIFE OF GOETHE," "THE PHYSIOLOGY OF COMMON LIFE,"
&c., &c.

NEW YORK:
HARPER & BROTHERS, PUBLISHERS,
FRANKLIN SQUARE.
1860.

CONTENTS.

CHAPTER I.

Omnipresence of Life.—The Microscope.—An Opalina and its Wonders.—The Uses of Cilia.—How our Lungs are protected from Dust and Filings.—Feeding without a Mouth or Stomach.—What is an Organ?—How a complex Organism arises.—Early Stages of a Frog and a Philosopher.—How the Plants feed.—Parasites of the Frog.—Metamorphoses and Migrations of Parasites.—Life within Life.—The budding of Animals.—A steady Bore.—Philosophy of the infinitely little ..... Page 9

CHAPTER II.

Ponds and Rock-pools. — Our necessary Tackle.— Wimbledon Common. — Early Memories.— Gnat Larvæ.— Entomostraca and their Paradoxes.—Races of Animals dispensing with the sterner Sex.—Insignificance of Males.—Volvox Globator: is it an Animal?—Plants swimming like Animals.—Animal Retrogressions.—The Dytiscus and its Larva.—The Dragon-fly Larva.—Mollusks and their Eggs.—Polypes, and how to find them.—A new Polype, *Hydra rubra*.—Nest-building Fish.—Contempt replaced by Reverence 39

CHAPTER III.

A garden Wall, and its Traces of past Life.—Not a Breath perishes.—A Bit of dry Moss and its Inhabitants.—The "Wheelbearers." — Resuscitation of Rotifers: drowned into Life.— Current Belief that Animals can be revived after complete Desiccation.—Experiments contradicting the Belief.—Spallanzani's Testimony.—Value of Biology as a Means of Culture.—Classification of Animals: the five great Types.—Criticism of Cuvier's Arrangement...................................... 59

CHAPTER IV.

An extinct Animal recognized by its Tooth: how came this to be possible?—The Task of Classification.—Artificial and natural Methods.—Linnæus, and his Baptism of the Animal Kingdom: his Scheme of Classification.—What is there underlying all true Classification?—The chief Groups.—What is a Species?—Restatement of the Question respecting the Fixity or Variability of Species.—The two Hypotheses.—Illustration drawn from the Romance Languages.—Caution to Disputants............Page 86

CHAPTER V.

Talking in Beetles.—Identity of Egyptian Animals with those now existing: Does this prove Fixity of Species?—Examination of the celebrated Argument of Species not having altered in four thousand Years.—Impossibility of distinguishing Species from Varieties.—The Affinities of Animals.—New Facts proving the Fertility of Hybrids.—The Hare and the Rabbit contrasted.—Doubts respecting the Development Hypothesis.—On Hypothesis in Natural History.—Pliny, and his Notion on the Formation of Pearls.—Are Pearls owing to a Disease of the Oyster?—Formation of the Shell; Origin of Pearls.—How the Chinese manufacture Pearls 107

CHAPTER VI.

Every Organism a Colony.—What is a Paradox?—An Organ is an independent Individual and a dependent one.—A Branch of Coral.—A Colony of Polypes.—The Siphonophora.—Universal Dependence.—Youthful Aspirings.—Our Interest in the Youth of great Men.—Genius and Labor.—Cuvier's College Life; his Appearance in Youth; his Arrival in Paris.—Cuvier and Geoffroy St. Hilaire.—Causes of Cuvier's Success.—One of his early Ambitions.—M. le Baron.—*Omnia vincit labor.*—Conclusion.. 128

STUDIES IN ANIMAL LIFE.

CHAPTER I.

Omnipresence of Life.—The Microscope.—An Opalina and its Wonders.—The Uses of Cilia.—How our Lungs are protected from Dust and Filings.—Feeding without a Mouth or Stomach. —What is an Organ?—How a complex Organism arises.— Early Stages of a Frog and a Philosopher.—How the Plants feed.—Parasites of the Frog.—Metamorphoses and Migrations of Parasites.—Life within Life.—The budding of Animals.— —A steady Bore.—Philosophy of the infinitely little.

COME with me, and lovingly study Nature, as she breathes, palpitates, and works under myriad forms of Life—forms unseen, unsuspected, or unheeded by the mass of ordinary men. Our course may be through park and meadow, garden and lane, over the swelling hills and spacious heaths, beside the running and sequestered streams, along the tawny coast, out on the dark and dangerous reefs, or under dripping caves and slippery ledges. It matters little where we go: every where—in the air above, the earth beneath, and waters under the earth—we are surrounded with Life. Avert your eyes a while from our human world, with its ceaseless anxieties, its noble sorrow, poignant, yet sublime, of conscious imperfection aspiring to higher states, and contemplate the calmer activities of that other world with which we are so mysteriously related. I hear you exclaim,

> "The proper study of mankind is man;"

nor will I pretend, as some enthusiastic students seem to think, that

> "The proper study of mankind is *cells;*"

but agreeing with you, that man is the noblest study, I would suggest that under the noblest there are other problems which we must not neglect. Man himself is imperfectly known, because the laws of universal Life are imperfectly known. His life forms but one grand illustration of Biology—the science of Life,* as he forms but the apex of the animal world.

Our studies here will be of Life, and chiefly of those minuter or obscurer forms, which seldom attract attention. In the air we breathe, in the water we drink, in the earth we tread on, Life is every where. Nature *lives:* every pore is bursting with Life; every death is only a new birth, every grave a cradle. And of this we know so little, think so little! Around us, above us, beneath us, that great mystic drama of creation is being enacted, and we will not even consent to be spectators! Unless animals are obviously useful or obviously hurtful

---

* The needful term Biology (from *Bios*, life, and *logos*, discourse) is now becoming generally adopted in England, as in Germany. It embraces all the separate sciences of Botany, Zoology, Comparative Anatomy, and Physiology.

to us, we disregard them. Yet they are not alien, but akin. The Life that stirs within us stirs within them. We are all "parts of one transcendent whole." The scales fall from our eyes when we think of this; it is as if a new sense had been vouchsafed to us, and we learn to look at Nature with a more intimate and personal love.

Life every where! The air is crowded with birds —beautiful, tender, intelligent birds—to whom life is a song and a thrilling anxiety, the anxiety of love. The air is swarming with insects—those little animated miracles. The waters are peopled with innumerable forms, from the animalcule, so small that one hundred and fifty millions of them would not weigh a grain, to the whale, so large that it seems an island as it sleeps upon the waves. The bed of the seas is alive with polypes, crabs, star-fishes, and with sand-numerous shell-animalcules. The rugged face of rocks is scarred by the silent boring of soft creatures, and blackened with countless mussels, barnacles, and limpets.

Life every where! on the earth, in the earth, crawling, creeping, burrowing, boring, leaping, running. If the sequestered coolness of the wood tempt us to saunter into its checkered shade, we are saluted by the murmurous din of insects, the twitter of birds, the scrambling of squirrels, the startled rush of unseen beasts, all telling how populous is this seeming solitude. If we pause before a tree, or shrub, or plant, our cursory and half-abstracted glance detects a colony of various inhabitants. We pluck a flower, and in its bosom we see many a charming insect busy at its appointed labor. We pick up a fallen leaf, and if nothing is visible on it, there is probably the trace of an insect larva hidden in its tissue, and awaiting there development. The drop of dew upon this leaf will probably contain its animals, visible under the microscope. This same microscope reveals that the *blood-rain* suddenly appearing on bread, and awakening superstitious terrors, is nothing but a collection of minute animals (*Monas prodigiosa*); and that the vast tracts of snow which are reddened in a single night owe their color to the marvelous rapidity in reproduction of a minute plant (*Protococcus nivalis*). The very mould which covers our cheese, our bread, our jam, or our ink, and disfigures our damp walls, is nothing but a collection of plants. The many-colored fire which sparkles on the surface of a summer sea at night, as the vessel plows her way, or which drips from the oars in lines of jeweled light, is produced by millions of minute animals.

Nor does the vast procession end here. Our very mother-earth is formed of the débris of life. Plants and animals which have been build up its solid fabric.* We dig downward thousands of feet below the surface, and discover with surprise the

---

* See EHRENBERG: *Microgeologie: das Erden und Felsen schaffende Wirken des unsichtbar kleinen selbstständigen Lebens auf der Erde.* 1854.

skeletons of strange, uncouth animals, which roamed the fens and struggled through the woods before man was. Our surprise is heightened when we learn that the very quarry itself is mainly composed of the skeletons of microscopic animals; the flints which grate beneath our carriage wheels are but the remains of countless skeletons. The Apennines and Cordilleras, the chalk cliffs so dear to homeward-nearing eyes—these are the pyramids of by-gone generations of atomies. Ages ago these tiny architects secreted the tiny shells which were their palaces; from the ruins of these palaces we build our Parthenons, our St. Peters, and our Louvres. So revolves the luminous orb of Life! Generations follow generations; and the Present becomes the matrix of the Future, as the Past was of the Present—the Life of one epoch forming the prelude to a higher Life.

When we have thus ranged air, earth, and water, finding every where a prodigality of living forms, visible and invisible, it might seem as if the survey were complete. And yet it is not so. Life cradles within Life. The bodies of animals are little worlds, having their own animals and plants. A celebrated Frenchman has published a thick octavo volume devoted to the classification and description of "The Plants which grow on Men and Animals;"* and many Germans have described the immense variety

* CHARLES ROBIN: *Histoire Naturelle des Végétaux Parasites qui croissent sur l'Homme et sur les Animaux Vivants.* 1853.

of animals which grow on and in men and animals; so that science can now boast of a parasitic Flora and Fauna. In the fluids and tissues, in the eye, in the liver, in the stomach, in the brain, in the muscles, parasites are found, and these parasites have often *their* parasites living in them!

We have thus taken a bird's-eye view of the field in which we may labor. It is truly inexhaustible. We may begin where we please, we shall never come to an end; our curiosity will never slacken.

> "And whosoe'er in youth
> Has through ambition of his soul given way
> To such desires, and grasp'd at such delights,
> Shall feel congenial stirrings late and long."

As a beginning, get a microscope. If you can not borrow, boldly buy one. Few purchases will yield you so much pleasure; and, while you are about it, do, if possible, get a good one. Spend as little money as you can on accessory apparatus and expensive fittings, but get a good stand and good glasses. Having got your instrument, bear in mind these two important trifles—work by daylight, seldom or never by lamplight; and keep the unoccupied eye *open*. With these precautions you may work daily for hours without serious fatigue to the eye.

Now where shall we begin? Any where will do. This dead frog, for example, that has already been made the subject of experiments, and is now awaiting the removal of its spinal cord, will serve us as a text from which profitable lessons may be drawn. We snip out a portion of its digestive tube, which, from its emptiness, seems to promise little; but a drop of the liquid we find in it is placed on a glass slide, covered with a small piece of very thin glass, and brought under the microscope. Now look. There are several things which might occupy your attention, but disregard them now to watch that animalcule which you observe swimming about. What is it? It is one of the largest of the Infusoria, and is named *Opalina*. When I call this an Infusorium I am using the language of text-books; but there seems to be a growing belief among zoologists that the Opalina is not an Infusorium, but the infantile condition of some worm (*Distoma?*). However, it will not grow into a mature worm as long as it inhabits the frog; it waits till some pike or bird has devoured the frog, and then, in the stomach of its new captor, it will develop into its mature form—then, and not till then. This surprises you. And well it may; but thereby hangs a tale, which to unfold—for the present, however, it must be postponed, because the Opalina itself needs all our notice.

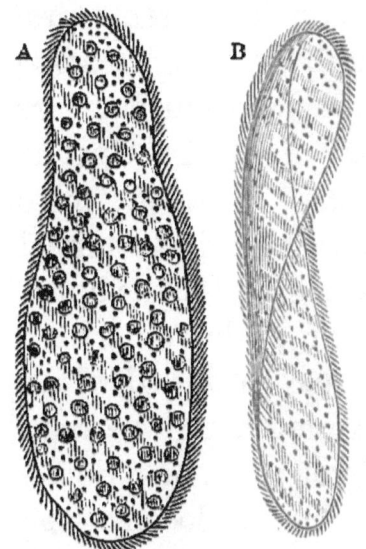

Fig. 1.—OPALINA RANARUM.
A, front view; B, side view—magnified.

Observe how transparent it is, and with what easy, undulating grace it swims about; yet this swimmer has no arms, no legs, no tail, no backbone to serve as a fulcrum to moving muscles—nay, it has no muscles to move with. 'Tis a creature of the most absolute abnegations—sans eyes, sans teeth, sans every thing; no, not sans every thing, for, as we look attentively, we see certain currents produced in the liquid, and, on applying a higher magnifying power, we detect how these currents are produced. All over the surface of the Opalina there are delicate hairs in incessant vibration; these are the *cilia*.* They lash the water, and the animal is propelled by their strokes, as a galley by its hundred oars. This is your first sight of that ciliary action of which you have so often read, and which you will henceforth find performing some important service in almost every animal you examine. Sometimes the cilia act as instruments of locomotion; sometimes as instruments of respiration, by continually renewing the current of water; sometimes as the means of drawing in food, for which purpose they surround the mouth, and by their incessant action produce a small whirlpool into which the food is sucked. An example of this is seen in the Vorticella. (Fig. 2.)

Having studied the action of these cilia in microscopic animals, you will be prepared to understand their office in your own organism. The lining

* From *cilium*, a hair.

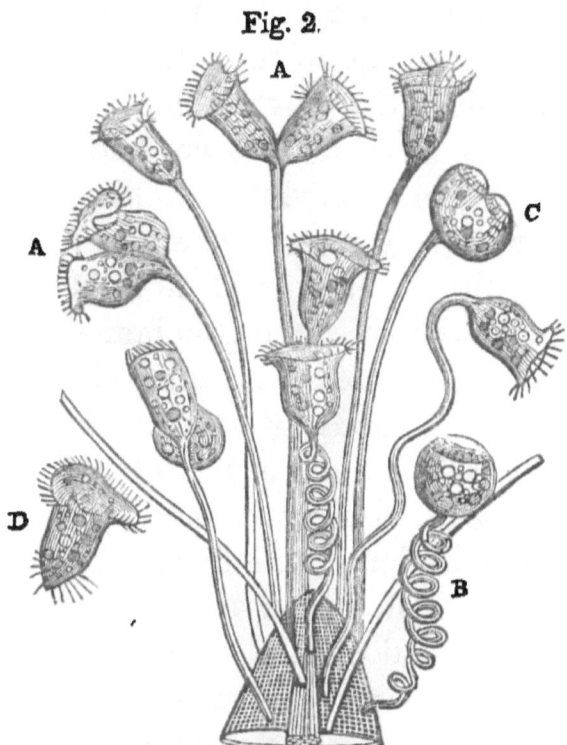

Group of Vorticella Nebulifera on a Stem of Weed, magnified.

A, one undergoing spontaneous division; B, another spirally retracted on its stalk; C, one with cilia retracted; D, a bud detached and swimming free.

membrane of your air-passages is covered with cilia, which may be observed by following the directions of Professor Sharpey, to whom science is indebted for a very exhaustive description of these organs. "To see them in motion, a portion of the ciliated mucous membrane may be taken from a recently-killed quadruped. The piece of membrane is to be folded with its free, or ciliated surface outward, placed on a slip of glass, with a little water or serum of blood, and covered with thin glass or mica. When it is now viewed with a power of 200 diameters or upward, a very obvious agitation will be perceived on the edge of the fold, and this appearance is caused by the moving cilia with which the surface of the membrane is covered. Being set close together, and moving simultaneously or in quick succession, the cilia, when in brisk action, give rise to the appearance of a bright transparent fringe along the fold of the membrane, agitated by such a rapid and incessant motion that the single threads which compose it can not be perceived. The motion here meant is that of the cilia themselves; but they also set in motion the adjoining fluid, driving it along the ciliated surface, as is indicated by the agitation of any little particles that may accidentally float in it. The fact of the conveyance of fluids and other matters along the ciliated surface, as well as the direction in which they are impelled, may also be made manifest by immersing the membrane in fluid, and dropping on it some finely-pulverized substance (such as charcoal in fine powder), which will be slowly but steadily carried along in a constant and determinate direction."*

It is an interesting fact, that while the direction in which the cilia propel fluids and particles is generally toward the interior of the organism, it is sometimes *reversed*, and, instead of beating the par-

---

* *Quain's Anatomy.* By SHARPEY and ELLIS. Sixth edition. I., p. lxxiii. See also SHARPEY's article *Cilia*, in the *Cyclopædia of Anatomy and Physiology.*

ticles inward, the cilia energetically beat them back if they attempt to enter. Fatal results would ensue if this were not so. Our air-passages would no longer protect the lungs from particles of sand, coal-dust, and filings flying about the atmosphere; on the contrary, the lashing hairs which cover the surface of these passages would catch up every particle, and drive it onward into the lungs. Fortunately for us, the direction of the cilia is reversed, and they act as vigilant janitors, driving back all vagrant particles with a stern "No admittance, *even on business!*" In vain does the whirlwind dash a column of dust in our faces—in vain does the air, darkened with coal-dust, impetuously rush up the nostrils; the air is allowed to pass on, but the dust is inexorably driven back. Were it not so, how could miners, millers, iron-workers, and all the modern Tubal Cains contrive to live in their loaded atmospheres? In a week their lungs would be choked up.

Perhaps you will tell me that this *is* the case—that manufacturers of iron and steel are very subject to consumption, and that there is a peculiar discoloration of the lungs which has often been observed in coal-miners examined after death.

Not being a physican, and not intending to trouble you with medical questions, I must still place before you three considerations, which will show how untenable this notion is. First, although consumption may be frequent among the Sheffield workmen, the cause is not to be sought in their breathing filings, but in the sedentary and unwholesome confinement incidental to their occupation. Miners and coal-heavers are not troubled with consumption. Moreover, if the filings were the cause, all the artisans would suffer, when all breathe the same atmosphere. Secondly, while it is true that discolored lungs have been observed in some miners, it has not been observed in all or in many; whereas it has been observed in men not miners, not exposed to any unusual amount of coal-dust. Thirdly, and most conclusively, experiment has shown that the coal-dust *can not* penetrate to the lungs. Claude Bernard, the brilliant experimenter, tied a bladder containing a quantity of powdered charcoal to the muzzle of a rabbit. Whenever the animal breathed, the powder within the bladder was seen to be agitated. Except during feeding-time the bladder was kept constantly on, so that the animal breathed only this dusty air. If the powder *could* have escaped the vigilance of the cilia and got into the lungs, this was a good occasion. But when the rabbit was killed and opened many days afterward, no powder whatever was found in the lungs or bronchial tubes; several patches were collected about the nostrils and throat, but the cilia had acted as a strainer, keeping all particles from the air-tubes.

The swimming apparatus of the Opalina has led us far away from the little animal who has been feeding while we have been lecturing. At the mention of feeding you naturally look for the food that is eaten, the mouth and stomach that eat. But I hinted just now that this ethereal creature dispenses with a stomach, as too gross for its nature, and of course, by a similar refinement, dispenses with a mouth. Indeed, it has no organs whatever except the cilia just spoken of. The same is true of several of the Infusoria, for you must know that naturalists no longer recognize the complex organization which Ehrenberg fancied he had detected in these microscopic beings. If it pains you to relinquish the piquant notion of a microscopic animalcule having a structure equal in complexity to that of the elephant, there will be ample compensation in the notion which replaces it, the notion of an ascending series of animal organisms, rising from the structureless *amœba* to the complex frame of a mammal. On a future occasion we shall see that, great as Ehrenberg's services have been, his *interpretations* of what he saw have one by one been replaced by truer notions. His immense class of Infusoria has been, and is constantly being, diminished; many of his animals turn out to be plants; many of them embryos of worms; and some of them belong to the same divisions of the animal kingdom as the oyster and the shrimp—that is to say, they range with the Mollusks and Crustaceans. In these, of course, there is a complex organization; but in the Infusoria, as now understood, the organization is extremely simple. No one now believes the clear spaces visible in their substance to be stomachs, as Ehrenberg believed; and the idea of the *Polygastrica*, or many-stomached Infusoria, is abandoned. No one now believes the colored specs to be eyes, because, not to mention the difficulty of conceiving eyes where there is no nervous system, it has been found that even the spores of some plants have these colored specs, and *they* are assuredly not eyes. If, then, we exclude the highly-organized *Rotifera*, or "Wheel Animalcules," which are genuine Crustacea, we may say that all Infusoria, whether they be the young of worms or not, are of very simple organization.

And this leads us to consider what biologists mean by an *organ:* it is a particular portion of the body set apart for the performance of some particular function. The whole process of development is this setting apart for special purposes. The starting-point of Life is a single cell—that is to say, a microscopic sac, filled with liquid and granules, and having within it a nucleus, or smaller sac. Paley has somewhere remarked that in the early stages there is no difference discernible between a frog and a philosopher. It is very true—truer than he conceived. In the earliest stage of all, both the Batrachian and the Philosopher are nothing but single cells, although the one cell will develop into an Aristotle or a Newton, and the other will get no higher than the cold, damp, croaking animal which boys will pelt, anatomists dissect, and Frenchmen eat. From the starting-point of a single cell this is the course taken: the cell divides itself into two, the two become four, the four eight, and so on, till a mass of cells is formed not unlike the shape of a mulberry. This mulberry-mass then becomes a sac, with double envelopes or walls; the inner wall, turned toward the yelk, or food, becomes the *assimilating* surface for the whole; the outer wall, turned toward the surrounding medium, becomes the surface which is to bring frog and philosopher into contact and relation with the external world—the Non-Ego, as the philosopher in after life will call it. Here we perceive the first grand "setting apart," or *differentiation*, has taken place; the embryo having an assimilating surface, which has little to do with the external world, and a sensitive, contractile surface, which has little to do with the preparation and transport of food. The embryo is no longer a mass of similar cells; it is already become dissimilar, *different*, as respects its inner and outer envelope. But these envelopes are at present uniform; one part of each is exactly like the rest. Let us, therefore, follow the history of Development, and we shall find that the inner wall gradually becomes unlike itself in various parts, and that certain organs, constituting a very complex apparatus of Digestion, Secretion, and Excretion, are all one by one wrought out of it by a series of metamorphoses or *differentiations*. The inner wall thus passes from a simple assimilating surface to a complex apparatus serving the functions of vegetative life.

Now glance at the outer wall: from it also various organs have gradually been wrought; it has developed into muscles, nerves, bones, organs of sense, and brain—all these from a simple homogeneous membrane!

With this bird's-eye view of the course of development you will be able to appreciate the grand law first clearly enunciated by Goethe and Von Baer as the law of animal life, namely, that development is always from the general to the special, from the simple to the complex, from the homogeneous to the heterogeneous, and this by a gradual series of *differentiations*.* Or, to put it into the music of our deeply meditative Tennyson,

> "All nature widens upward. Evermore
> The simpler essence lower lies:
> More complex is more perfect—owning more
> Discourse, more widely wise."

You are now familiarized with the words "differentiation" and "development," so often met with in modern writers, and have gained a distinct idea of what an "organ" is, so that, on hearing of an animal without organs, you will at once conclude that in such an animal there has been no setting apart of any portion of the body for special purposes, but that all parts serve all purposes indis-

---

* GOETHE: *Zur Morphologie*, 1807. VON BAER: *Zur Entwickelungsgeschichte*, 1828. Part I., p. 158.

criminately. Here is our Opalina, for example, without mouth, or stomach, or any other organ. It is an assimilating surface in every part; in every part a breathing, sensitive surface. Living on liquid food, it does not need a mouth to seize, or a stomach to digest such food. The liquid, or gas, passes through the Opalina's delicate skin by a process which is called *endosmosis;* it there serves as food; and the refuse passes out again by a similar process, called *exosmosis*. This is the way in which many animals and all plants are nourished. The cell at the end of a rootlet, which the plant sends burrowing through the earth, has no mouth to seize, no open pores to admit the liquid that it needs; nevertheless, the liquid passes into the cell through its delicate cell-wall, and passes from this cell to *other* cells upward from the rootlet to the bud. It is in this way, also, that the Opalina feeds: it is all-mouth, no-mouth; all-stomach, no-stomach. Every part of its body performs the functions which in more complex animals are performed by organs specially set apart. It feeds without mouth, breathes without lungs, and moves without muscles.

The Opalina, as I said, is a parasite. It may be found in various animals, and almost always in the frog. You will perhaps ask why it should be considered a parasite? why may it not have been swallowed by the frog in a gulp of water? Certainly nothing would have been easier. But, to remove your doubts, I open the skull of this frog, and carefully remove a drop of the liquid found inside, which, on being brought under a microscope, we shall most probably find containing some animalcules, especially those named *Monads*. These were not swallowed. They live in the cerebro-spinal fluid, as the Opalina lives in the digestive tube. Nay, if we extend our researches, we shall find that various organs have their various parasites. Here, for instance, is a parasitic worm from the frog's bladder. Place it under the microscope with a high power, and behold! It is called *Polystomum*—many-mouthed, or, more properly, many-suckered. You are looking at the under side, and will observe six large suckers with their starlike clasps (*e*), and the horny instrument (*f*) with which the animal bores its way. At *a* there is another sucker, which serves also as a mouth; at *b* you perceive the rudiment of a gullet, and at *d* the reproductive organs. But pay attention to the pretty branchings of the digestive tube (*c*), which ramifies through the body like a blood-vessel.

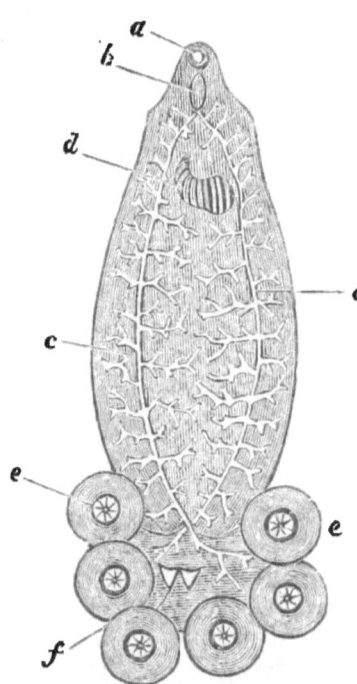

Fig. 3.—Polystomum Integerrimum, magnified.

This arrangement of the digestive tube is found

in many animals, and is often mistaken for a system of blood-vessels. In one sense this is correct, for these branching tubes are carriers of nutriment, and the only circulating vessels such animals possess; but the nutriment is *chyme*, not blood: these simple animals have not arrived at the dignity of blood, which is a higher elaboration of the food, fitted for higher organisms.

Thus may our frog, besides its own marvels, afford us many "authentic tidings of invisible things," and is itself a little colony of life. Nature is economic as well as prodigal of space. She fills the illimitable heavens with planetary and starry grandeurs, and the tiny atoms moving over the crust of earth she makes the homes of the infinitely little. Far as the mightiest telescope can reach, it detects worlds in clusters, like pebbles on the shore of infinitude; deep as the microscope can penetrate, it detects life within life, generation within generation, as if the very universe itself were not vast enough for the energies of life!

That phrase, generation within generation, was not a careless phrase; it is exact. Take the tiny insect (*Aphis*) which, with its companions, crowds your rose-tree; open it, in a solution of sugar-water, under your microscope, and you will find in it a young insect nearly formed; open that young insect with care, and you will find in it, also, another young one, less advanced in its development, but perfectly recognizable to the experienced eye; and beside this embryo you will find many eggs, which would in time become insects!

Or take that lazy water-snail (*Paludina vivipara*), first made known to science by the great Swammerdamm, the incarnation of patience and exactness, and you will find, as he found, forty or fifty young snails in various stages of development; and you will also find, as he found, some tiny worms, which, if you cut them open, will suffer three or four infusoria to escape from the opening.* In your astonishment you will ask, Where is this to end?

The observation recorded by Swammerdamm, like so many others of this noble worker, fell into neglect, but modern investigators have made it the starting-point of a very curious inquiry. The worms he found within the snail are now called *Cercaria sacs*, because they contain the *Cercariæ*, once classed as Infusoria, and which are now known to be the early forms of parasitic worms inhabiting the digestive tube and other cavities of higher animals. These *Cercariæ* have vigorous tails, with which they swim through the water like tadpoles, and, like tadpoles, they lose their tails in after life. But how, think you, did these sacs containing *Cercariæ* get into the water-snails? "By spontaneous generation," formerly said the upholders of that hypothesis, and those who condemned the hypothesis were forced to admit they had no better explanation. It was a mystery which they preferred leaving unex-

* SWAMMERDAMM. *Bibel der Natur*, p. 75-77.

plained rather than fly to spontaneous generation. And they were right. The mystery has at length been cleared up.* I will endeavor to bring together the scattered details, and narrate the curious story.

Under the eyelids of geese and ducks may be constantly found a parasitic worm (of the *Trematode* order), which naturalists have christened *Monostomum mutabile*—Single-mouth, Changeable. This worm brings forth living young, in the likeness of active Infusoria, which, being covered with cilia, swim about in the water as we saw the Opalina swim. Here is a portrait of one.

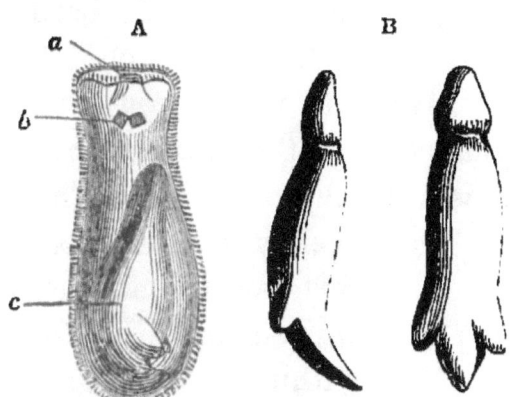

Fig. 4.—A, Embryo of Monostomum Mutabile. B, Cercaria sac, just set free.
a, mouth; *b*, pigment spots; *c*, sac—magnified.

Each of these animalcules develops a sac in its interior. The sac you may notice in the engraving.

* By Von Siebold. See his interesting work, *Ueber die Band- und-Blasenwürmer*. It has been translated by Huxley, and appended to the translation of Kuechenmeister *on Parasites*, published by the Sydenham Society.

Having managed to get into the body of the water-snail, the animalcule's part in the drama is at an end. It dies, and in dying liberates the sac, which is very comfortably housed and fed by the snail. If you examine this sac (Fig. 5), you will observe that it has a mouth and digestive tube, and is, therefore, very far from being, what its name imports, a mere receptacle; it is an independent animal, and lives an independent life. It feeds generously on the juices of the snail, and, having fed, thinks generously of the coming generations. It was born inside the animalcule; why should it not in turn give birth to children of its own? To found a dynasty, to scatter progeny over the bounteous earth, is a worthy ambition. The mysterious agency of reproduction begins in this sac-animal, and in a short while a brood of *Cercariæ* move within it. The sac bursts, and the brood escapes. But how is this? The children are by no means the "very image" of their parent. They are not sacs, nor in the least resembling sacs, as you see (Fig. 6).

They have tails, and suckers, and sharp boring instruments,

Fig. 5.—CERCARIA SAC.

A, mouth; B, digestive tube; C, a cercaria newly formed: four others are seen in different stages—magnified.

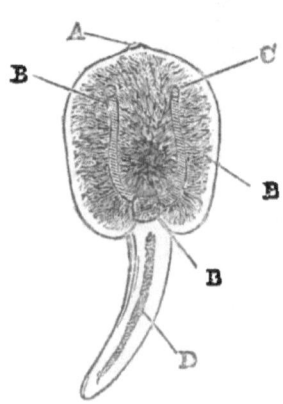

Fig. 6.—CERCARIA DEVELOPED.

A, mouth; B, B, B, excretory organ; C, pigment spots; D, tail.

with other organs which their parent was without. To look at them you would as soon suspect a shrimp to be the progeny of an oyster, as these to be the progeny of the sac-animal. And what makes the paradox more paradoxical is, that not only are the *Cercariæ* unlike their parent, but their parent was equally unlike its parent, the embryo of *Monostomum* (compare Fig. 4). However, if we pursue this family history, we shall find the genealogy rights itself at last, and that this Cercaria will develop in the body of some bird into a *Monostomum mutabile* like its ancestor. Thus the worm produces an animalcule, which produces a sac-animal, which produces a Cercaria, which becomes a worm exactly resembling its great-grandfather.

One peculiarity in this history is, that while the *Monostomum* produces its young in the usual way, the two *intermediate* forms are produced by a process of budding analogous to that observed in plants. Plants, as you know, are reproduced in two ways —from the seed and from the bud. For seed-reproduction peculiar organs are necessary; for bud-reproduction there is no such differentiation needed: it is simply an outgrowth. The same is true of many animals; they also bud like plants, and produce seeds (eggs) like plants. I have elsewhere argued that the two processes are essentially identical, and that both are but special forms of growth.* Not, however, to discuss so abstruse a

* *Seaside Studies*, p. 308 *et sq*.

question here, let us merely note that the Monostomum, into which the Cercaria will develop, produces eggs, from which young will issue; the second generation is not produced from eggs, but by internal budding; the third generation is likewise budded internally, but it, on acquiring maturity, will produce eggs. For this maturity, it is indispensable that the Cercaria should be swallowed by some bird or animal; only in the digestive tube can it acquire its producing condition. How is it to get there? The ways are many; let us witness one:

In this watch-glass of water we have several *Cercariæ* swimming about. To them we add three or four of those darting, twittering insects which you have seen in every vase of pond-water, and have learned to be the larvæ or early forms of the *Ephemeron*. The *Cercariæ* cease flapping the water with their impatient tails, and commence a severe scrutiny of the strangers. When Odry, in the riotous farce *Les Saltimbanques*, finds a portmanteau, he exclaims, "*Un malle! ce doit être à moi!*" ("Surely this *must* belong to me!") This seems to be the theory of property adopted by the Cercaria: "An insect! surely this belongs to me!" Accordingly every one begins creeping over the bodies of the Ephemera, giving an interrogatory poke with the spine, which will pierce the first soft place it can detect. Between the segments of the insect's armor a soft and pierceable spot is found; and now, lads, to work! Onward they bore, never relaxing in their efforts till a hole is made large enough for them to slip in by elongating their bodies. Once in, they dismiss their tails as useless appendages, and begin what is called the process of *encysting*—that is, of rolling themselves up into a ball, and secreting a mucus from their surface which hardens round them like a shell. Thus they remain snugly ensconced in the body of the insect, which in time develops into a fly, hovers over the pond, and is swallowed by some bird. The fly is digested, and the liberated Cercaria finds itself in comfortable quarters, its shell is broken, and its progress to maturity is rapid.

Von Siebold's description of another form of emigration he has observed in parasites will be read with interest. "For a long time," he says, "the origin of the thread-worm, known as *Filaria insectorum*, that lives in the cavity of the bodies of adult and larval insects, could not be accounted for. Shut up within the abdominal cavity of caterpillars, grasshoppers, beetles, and other insects, these parasites were supposed to originate by spontaneous generation under the influence of wet weather or from decayed food. Helminthologists (students of parasitic worms) were obliged to content themselves with this explanation, since they were unable to find a better. Those who dissected these thread-worms, and submitted them to a careful inspection, could not deny the probability, since it was clear that they contained no trace of sexual organs. But, on directing my attention to these entozoa, I became aware of the fact that they were not true *Filariæ* at all, but belonged to a peculiar family of thread-worms, embracing the genera of *Gordius* and *Mermis*. Furthermore, I convinced myself that these parasites wander away when full grown, boring their way from within through any soft place in the body of their host, and creeping out through the opening. These parasites do not emigrate because they are uneasy, or because the caterpillar is sickly, but from that same internal necessity which constrains the horsefly to leave the stomach of the horse where he has been reared, or which moves the gadfly to work its way out through the skin of the ox. The larvæ of both these insects creep forth in order to become chrysalises, and thence to proceed to their higher and perfect condition. I have demonstrated that the perfect, full-grown, but *sexless* thread-worms of insects are in like manner moved by their desire to wander out of their previous homes in order to enter upon a new period of their lives, which ends in the development of their sex. As they leave the bodies of their hosts, they fall to the ground and crawl away into the deeper and moister parts of the soil. Thread-worms found in the damp earth, in digging up gardens and cutting ditches, have often been brought to me which presented no external distinctions from the thread-worms of insects. This suggested to me that the wandering thread-worms of insects might instinctively bury themselves in damp ground, and I therefore instituted a series of experiments by placing the newly-emigrated worms in flower-pots filled with damp earth. To my delight, I soon perceived that they began to bore with their heads into the earth, and by degrees drew themselves entirely in. For many months I kept the earth in the flower-pots moderately moist, and, on examining the worms from time to time, I found they had gradually attained their sex-development, and eggs were deposited in hundreds. Toward the conclusion of winter I could succeed in detecting the commencing development of the embryos in these eggs. By the end of spring they were fully formed, and many of them, having left their shells, were to be seen creeping about the earth. I now conjectured that these young worms would be impelled by their instincts to pursue a parasitic existence, and to seek out an animal to inhabit and to grow to maturity in; and it seemed not improbable that the brood I had reared would, like their parents, thrive best in the caterpillar. In order, therefore, to induce my young brood to immigrate, I procured a number of very small caterpillars, which the first spring sunshine had just called into life. For the purpose of my experiment, I filled a watch-glass with damp earth, taking it from among the flower-pots where the thread-worms had wintered. Upon this I placed several of the young caterpillars." The result was as he expected; the caterpillars were soon bored into by the worms, and served them at once as food and home.*

Frogs and parasites, worms and infusoria—are these worth the attention of a serious man? They have a less imposing appearance than planets and asteroids I admit, but they are nearer to us, and admit of being more intimately known, and, because they are thus accessible, they become more important to us. The life that stirs within us is also the life within them. It is for this reason, as I said at the outset, that, although man's noblest study must always be man, there are other studies less noble, yet not therefore ignoble, which must be pursued, even if only with a view to the perfection of the noblest. Many men, and these not always the ignorant, whose scorn of what they do not understand is always ready, despise the labors which do not obviously and directly tend to moral or political advancement. Others there are who, fascinated by the grandeur of Astronomy and Geology, or by the immediate practical results of Physics and Chemistry, disregard all microscopic research as little better than dilettante curiosity. But I can not think any serious study is without its serious value to the human race; and I know that the great problem of Life can never be solved while we are in ignorance of its simpler forms. Nor can any thing be more unwise than the attempt to limit the sphere

---

* Von Siebold: *Ueber Band-und-Blasenwürmer.* Translated by Huxley.

of human inquiry, especially by applying the test of immediate utility. All truths are related; and, however remote from our daily needs some particular truth may seem, the time will surely come when its value will be felt. To the majority of our countrymen during the Revolution, when the conduct of James seemed of incalculable importance, there would have seemed something ludicrously absurd in the assertion that the newly-discovered differential calculus was infinitely more important to England and to Europe than the fate of all the dynasties; and few things could have seemed more remote from any useful end than this product of mathematical genius; yet it is now clear to every one that the conduct of James was supremely insignificant in comparison with this discovery. I do not say that men were unwise to throw themselves body and soul into the Revolution; I only say they would have been unwise to condemn the researches of mathematicians.

Let all who have a longing to study Nature in any of her manifold aspects do so without regard to the sneers or objections of men whose tastes and faculties are directed elsewhere. From the illumination of many minds on many points Truth must finally emerge. Man is, in Bacon's noble phrase, the minister and interpreter of Nature; let him be careful lest he suffer this ministry to sink into a priesthood, and this interpretation to degenerate into an immovable dogma. The suggestions of apathy and the prejudices of ignorance have at all times inspired the wish to close the temple against new comers. Let us be vigilant against such suggestions, and keep the door of the temple ever open.

CHAPTER II.

Ponds and Rock-pools. — Our necessary Tackle. — Wimbledon Common. — Early Memories. — Gnat Larvæ. — Entomostraca and their Paradoxes. — Races of Animals dispensing with the sterner Sex. — Insignificance of Males. — Volvox Globator: is it an Animal? — Plants swimming like Animals. — Animal Retrogressions. — The Dytiscus and its Larva. — The Dragon-fly Larva. — Mollusks and their Eggs. — Polypes, and how to find them. — A new Polype, *Hydra rubra*. — Nest-building Fish. — Contempt replaced by Reverence.

THE day is bright with a late autumn sun; the sky is clear with a keen autumn wind, which lashes our blood into a canter as we press against it, and the cantering blood sets the thoughts into hurrying excitement. Wimbledon Common is not far off; its five thousand acres of undulating heather, furze, and fern tempt us across it, health streaming in at every step as we snuff the keen breeze. We are tempted also to bring net and wide-mouthed jar, to ransack its many ponds for visible and invisible wonders.

Ponds, indeed, are not so rich and lovely as rock-pools; the heath is less alluring than the coast— our dear-loved coast, with its gleaming mystery, the sea, and its sweeps of sand, its reefs, its dripping boulders. I admit the comparative inferiority of ponds, but, you see, we are not near the coast, and the heath is close at hand. Nay, if the case were otherwise, I should object to dwarfing comparisons. It argues a pitiful thinness of nature (and the majority in this respect *are* lean) when present excellence is depreciated because some greater excellence is to be found elsewhere. We are not elsewhere; we must do the best we can with what is here. Because ours is not the Elizabethan age, shall we express no reverence for our great men, but reserve it for Shakspeare, Bacon, and Raleigh, whose traditional renown must overshadow our contemporaries? Not so. To each age its honor. Let us be thankful for all greatness, past or present, and never speak slightingly of noble work or honest endeavor because it is not, or we choose to say it is not, equal to something else. No comparisons, then, I beg. If I said ponds were finer than rock-pools, you might demur; but I only say ponds are excellent things, let us dabble in them; ponds are rich in wonders, let us enjoy them.

And, first, we must look to our tackle. It is extremely simple. A landing-net, lined with muslin; a wide-mouthed glass jar, say a foot high and six inches in diameter, but the size optional, with a bit of string tied under the lip, and forming a loop over the top, to serve as a handle, which will let the jar swing without spilling the water; a camel's-hair brush; a quinine bottle, or any wide-mouthed phial, for worms and tiny animals which you desire to keep separated from the dangers and confusions of the larger jar; and when to these a pocket lens is added, our equipment is complete.

As we emerge upon the common and tread its springy heather, what a wild wind dashes the hair into our eyes, and the blood into our cheeks! and what a fine sweep of horizon lies before us! The lingering splendors and the beautiful decays of autumn vary the scene, and touch it with a certain pensive charm. The ferns mingle harmoniously their rich browns with the dark green of the furze, now robbed of its golden summer glory, but still pleasant to the eye and exquisite to memory. The gaunt wind-mill on the rising ground is stretching its stiff, starred arms into the silent air, a landmark for the wanderer—a land-mark, too, for the wandering mind, since it serves to recall the dim early feelings and sweet broken associations of a childhood when we gazed at it with awe, and listened to the rushing of its mighty arms. Ah! well may the mind with the sweet insistance of sadness linger on those scenes of the irrecoverable past, and try, by lingering there, to feel that it is not wholly lost, wholly irrecoverable, vanished forever from the Life which, as these decays of autumn and these changing trees too feelingly remind us, is gliding away, leaving our cherished ambitions still unfulfilled, and our deeper affections still but half expressed. The vanishing visions of elapsing life bring with them thoughts which lie too deep for tears, and this wind-mill recalls such visions by the subtle laws of association. Let us go toward it, and stand once more under its shadow. See the intelligent and tailless sheep-dog which bounds out at our approach, eager and minatory; now his quick eye at once recognizes that we are neither tramps nor thieves, and he ceases barking to commence a lively interchange of sniffs and amenities with our Pug, who seems also glad of a passing interchange of commonplace remarks. While these dogs travel over each other's minds, let us sun ourselves upon this bench, and look down on the embrowned valley, with its gipsy encampment, or abroad on the purple Surrey hills, or the varied-tinted trees of Combe Wood and Richmond Park. There are not many such prospects so near London. But, in spite of the sun, we must not linger here: the wind is much too analytical in its remarks; and, moreover, we came out to hunt.

Here is a pond with a mantling surface of green promise. Dip the jar into the water. Hold it now up to the light, and you will see an immense variety of tiny animals swimming about. Some are large enough to be recognized at once; others require a pocket lens, unless familiarity has already enabled you to *infer* the forms you can not distinctly *see*. Here (Fig. 7) are two larvæ (or grubs) of the common gnat. That large-headed fellow (A) bobbing about with such grotesque movements is very near the last stage of his metamorphosis, and to-morrow, or the next day, you may see him

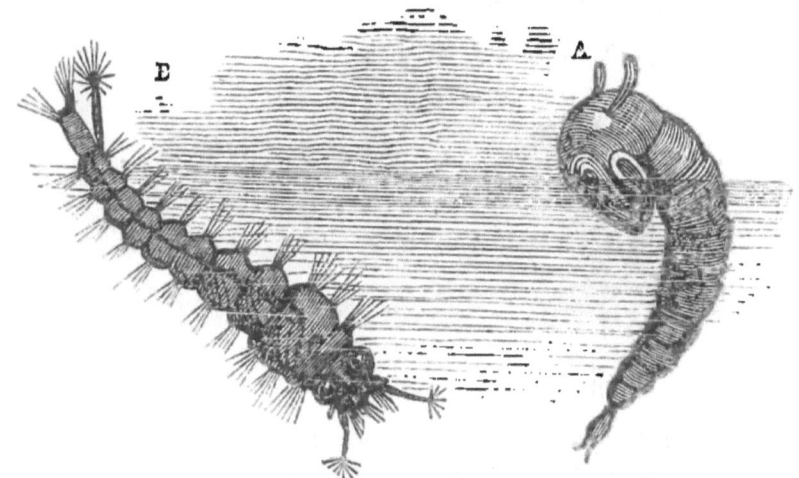

Fig. 7.—LARVÆ OF THE GNAT in two different stages of development (magnified).

cast aside this mask (*larva* means a mask), and emerge a perfect insect. The other (B) is in a much less matured condition, but leads an active predatory life, jerking through the water, and fastening to the stems of weed or sides of the jar by means of the tiny hooks at the end of its tail. The hairy appendage forming the angle is not another tail, but a breathing apparatus.

Observe, also, those grotesque *Entomostraca*,* popularly called "water-fleas," although, as you perceive, they have little resemblance in form or manners to our familiar (somewhat *too* familiar) bed-fellows. This (Fig. 8) is a *Cyclops*, with only one eye in the centre of its forehead, and carrying two sacs, filled with eggs, like panniers. You ob-

* *Entomostraca* (from *entomos*, an insect, and *ostracon*, a shell) are not really insects, but belong to the same large group of animals as the lobster, the crab, or the shrimp—*i. e.*, crustaceans.

serve he has no legs; or, rather, legs and arms are

Fig. 8.—CYCLOPS.
a, large antennæ; *b*, smaller do.; *c*, egg-sacs (magnified).

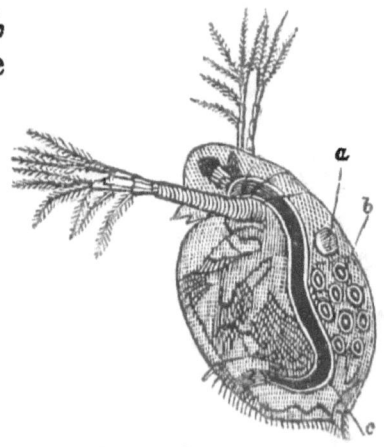

Fig. 9.—DAPHNIA.
a, pulsatile sac, or heart; *b*, eggs; *c*, digestive tube (magnified).

hoisted up to the head, and become antennæ (or feelers). Here (Fig. 9) is a *Daphnia*, grotesque enough, throwing up his arms in astonished awkwardness, and keeping his legs actively at work inside the shell—as respirators, in fact. Here (Fig. 10) is a *Eurycercus*, less grotesque, and with a much smaller eye. Talking of eyes, there is one of these Entomostraca, named *Polyphemus*, whose head is all eye; and another, named *Caligus*, who has no head at all. Other paradoxes and wonders are presented by this interesting group of animals;* but they all sink

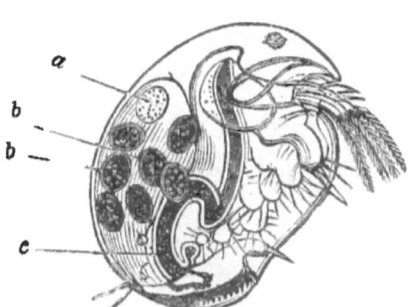

Fig. 10.—EURYCERCUS.
a, heart; *b*, eggs; *c*, digestive tube (magnified).

* The student will find ample information in BAIRD's *British Entomostraca*, published by the Ray Society.

into insignificance beside the paradox of the amazonian entomostracon, the *Apus*—a race which dispenses with masculine services altogether, a race of which there are no males!

I well remember the pleasant evening on which I first made the personal acquaintance of this amazon. It was at Munich, and in the house of a celebrated naturalist, in whose garden an agreeable assemblage of poets, professors, and their wives sauntered in the light of a setting sun, breaking up into groups and *têtes-à-têtes*, to re-form into larger groups. We had taken coffee under the branching coolness of trees, and were now loitering through the brief interval till supper. Our host had just returned from an expedition of some fifty miles to a particular pond, known to be inhabited by the Apus. He had made this journey because the race, although prolific, is rare, and is not to be found in every spot. For three successsive years had he gone to the same pond in quest of the male; but no male was to be found among thousands of egg-bearing females, some of which he had brought away with him, and was showing us. We were amused to see them swimming about, sometimes on their backs, using their long oars, sometimes floating, but always incessantly agitating the water with their ten pairs of breathing legs; and the ladies, gathered round the jar, were hugely elated at the idea of animals getting rid altogether of the sterner sex—clearly a useless encumbrance in the scheme of things!

The fact that no male Apus has yet been found is not without precedent. Léon Dufour, the celebrated entomologist, declares that he never found the male of the gall insect (*Diplolepis gallæ tinctoriæ*), though he has examined thousands: they were all females, and bore well-developed eggs on emerging from the gall-nut in which their infancy had passed. In two other species of gall insect—*Cynips divisa* and *Cynips folii*—Hartig says he was unable to find a male; and he examined about thirteen thousand. Brongniart never found the male of another entomostracon (*Limnadia gigas*), nor could Jurine find that of our *Polyphemus*. These negatives prove, at least, that if the males exist at all, they must be excessively rare, and their services can be dispensed with; a conclusion which becomes acceptable when we learn that bees, plant-lice (*Aphides*), and our grotesque friend *Daphnia* (Fig. 9) lay eggs which may be reared apart, will develop into females, and these will produce eggs which will in turn produce other females, and so on, generation after generation, although each animal be reared in a vessel apart from all others.

While on this subject, I can not forbear making a reflection. It must be confessed that our sex cuts but a poor figure in some great families. If the male is in some families grander, fiercer, more splendid, and more highly endowed than the female, this occasional superiority is more than counterbalanced by the still greater inferiority of the sex in other families. The male is often but a contemptible partner, puny in size, insignificant in powers, stinted even of a due allowance of organs. If the peacock and the pheasant swagger in greater splendor, what a pitiful creature is the male falcon!—no falconer will look at him. And what is the drone compared with the queen bee, or even with the workers? What figure does the male spider make beside his large and irascible female, who not unfrequently eats him? Nay, worse than this, what can be said for the male Rotifer, the male Barnacle, the male Lernæa—gentlemen who can not even boast of a perfect digestive apparatus, sometimes not of a digestive organ at all? Nor is this meagreness confined to the digestive system only. In some cases, as in some male Rotifers, the usual organs of sense and locomotion are wanting;* and in a parasitic Lernæa, the degradation is moral as well as physical: the female lives in the gills of a fish, sucking its juices, and the ignoble husband lives as a parasite upon her!

But this digression is becoming humiliating, and meanwhile our hands are getting benumbed with cold. In spite of that, I hold the jar up to the light, and make a background of my forefingers, to throw into relief some of the transparent animals. Look at those green crystal spheres sailing along

---

* Compare GEGENBAUR: *Grundzüge der vergleichende Anatomie*, 1859, p. 229 *und* 269; also LEYDIG *über Hydatina senta*, in *Müller's Archiv*, 1857, p. 411.

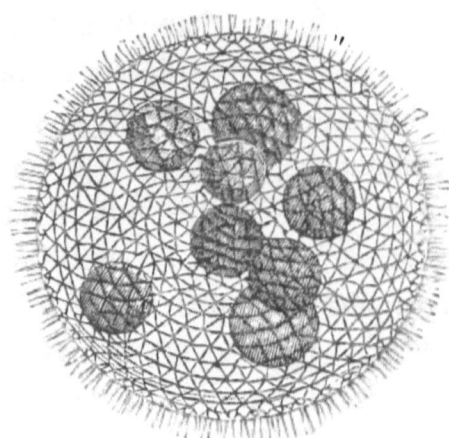

Fig. 11.—VOLVOX GLOBATOR, with eight volvoces inclosed (magnified).

with slow revolving motion, like planets revolving through space, except that their orbits are more eccentric. Each of these spheres is a *Volvox globator*. Under the microscope it looks like a crystalline sphere, studded with bright green specks, from each of which arise two cilia (hairs), serving as oars to row the animal through the water. The specks are united by a delicate net-work, which is not always visible, however. Inside this sphere is a fluid, in which several dark green smaller spheres are seen revolving, as the parent sphere revolved in the water. Press this Volvox gently under your compressorium, or between the two pieces of glass, and you will see these internal spheres, when duly magnified, disclose themselves as identical with their parent; and inside them smaller Volvoces are seen. This is one of the many illustrations of life within life, of which something was said in the last chapter.

Nor is this all. Those bright green specks which stud the surface, if examined with high powers, will turn out to be, not specks, but animals,* and, as Ehrenberg believes (though the belief is but little shared), highly organized animals, possessing a mouth, many stomachs, and an eye. It is right to add that not only are microscopists at variance with Ehrenberg on the supposed organization of these specks, but the majority deny that the Volvox itself is an animal. Von Siebold in Germany, and Professor George Busk and Professor Williamson in England, have argued with so much force against the animal nature of the Volvox, which they call a plant, that in most modern works you will find this opinion adopted. But the latest of the eminent authorities on the subject of Infusoria, in his magnificent work just published, returns to the old idea that the Volvox is an animal after all, although of very simple organization.†

The dispute may perhaps excite your surprise. You are perplexed at the idea of a plant (if plant it be) moving about, swimming with all the vigor and dexterity of an animal, and swimming by means of animal organs, the cilia. But this difficulty is one of our own creation. We first employ the word

* To avoid the equivoque of calling the parts of an animal, which are capable of independent existence, by the same term as the whole mass, we may adopt Huxley's suggestion, and call all such individual parts *zöoids* instead of animals. Dugès suggested *zöonites* in the same sense.—*Sur la Conformité Organique*, p. 13.

† Stein: *Der Organismus der Infusionsthiere*, 1859, p. 36-38.

plant to designate a vast group of objects which have no powers of locomotion, and then ask, with triumph, How can a plant move? But we have only to enlarge our knowledge of plant-life to see that locomotion is not absolutely excluded from it; for many of the simpler plants—Confervæ and Algæ—can and do move spontaneously in the early stages of their existence: they escape from their parents as free swimming rovers, and do not settle into solid and sober respectability till later in life. In their roving condition they are called, improperly enough, "zoospores,"* and once gave rise to the opinion that they were animals in infancy, and became degraded into plants as their growth went on. But locomotion is no true mark of animal-nature, neither is fixture to one spot the true mark of plant-nature. Many animals (Polypes, Polyzoa, Barnacles, Mussels, etc.), after passing a vagabond youth, "settle" once and forever in maturer age, and then become as fixed as plants. Nay, human animals not unfrequently exhibit a somewhat similar metempsychosis, and make up for the fitful capriciousness of wandering youth by the steady severity of their application to business when width of waistcoat and smoothness of cranium suggest a sense of their responsibilities.

Whether this loss of locomotion is to be regarded as a retrogression on the part of the plant or animal which becomes fixed, may be questioned;

---

* Zoospores, from *zoon*, an animal, and *sporos*, a seed.

but there are curious indications of positive retrogression from a higher standard in the metamorphoses of some animals. Thus the beautiful marine worm *Terebella*, which secretes a tube for itself, and lives in it, fixed to the rock or oyster-shell, has in early life a distinct head, eyes, and feelers; but in growing to maturity it loses all trace of head, eyes, and even of feelers, unless the beautiful tuft of streaming threads which it waves in the water be considered as replacing the feelers. There are the Barnacles, too, which in the first stage of their existence have three pairs of legs, a very simple single eye, and a mouth furnished with a proboscis. In the second stage they have six pairs of legs, two compound eyes complex in structure, two feelers, but *no mouth*. In the third, or final stage, their legs are transformed into prehensile organs, they have recovered a mouth, but have lost their feelers, and their two complex eyes are degraded to a single and very simple eye-spot.

But, to break up these digressions, let us try a sweep with our net. We skim it along the surface, and draw up a quantity of duckweed, dead leaves, bits of stick, and masses of green thread of great fineness, called Conferva by botanists. The water runs away, and we turn over the mass. Here is a fine water-beetle, called the "Water-tiger," from its ferocity (Fig. 12). You would hardly suspect that the slim, big-headed, long-tailed Water-tiger would grow into the squat, small-headed, tailless beetle:

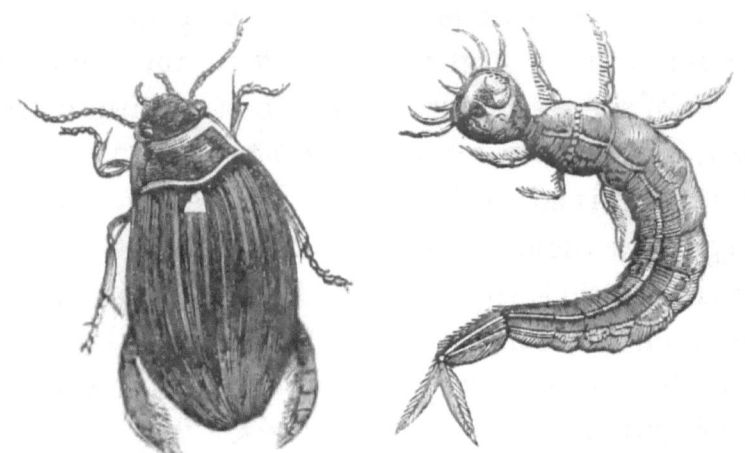

Fig. 12.—WATER BEETLE and its larva.

nor would you imagine that this Water-tiger would be so "high fantastical" as to breathe by his tail. Yet he does both, as you will find if you watch him in your aquarium.

Continuing our search, we light upon the fat, sluggish, ungraceful larva of the graceful and brilliant Dragon-fly, the falcon of insects (Fig. 13). He is useful for dissection, so pop him in. Among the dead leaves you per-

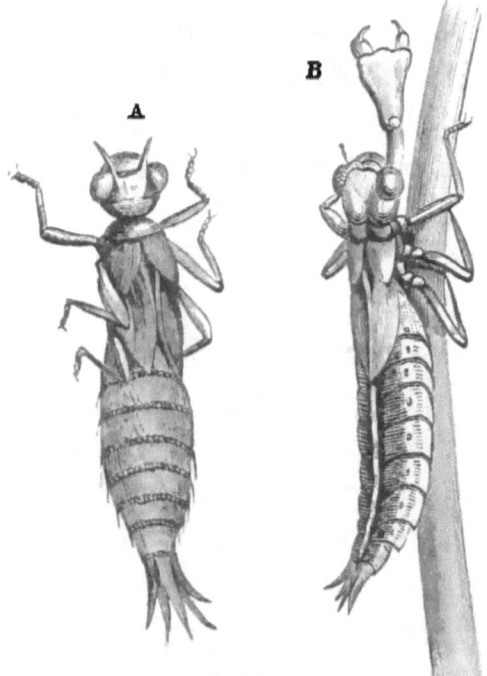

Fig. 13.—DRAGON-FLY LARVÆ:
A, ordinary aspect; B, with the huge nipper-like jaw extended.

ceive several small leeches, and flat oval *Planariæ*, white and brown; and here also is a jelly-like mass, of pale yellow color, which we know to be a mass of eggs deposited by some shell-fish; and, as there are few objects of greater interest than an egg in course of development, we pop the mass in. Here (Fig. 14) are two mollusks, *Lim-*

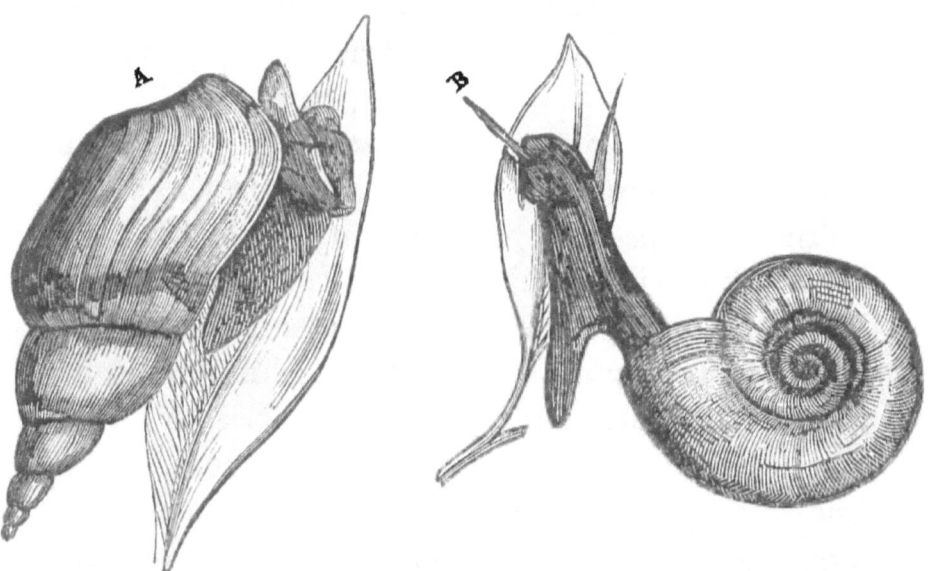

Fig. 14.—A, Limnæus Stagnalis, or Water-snail.   B, Planorbis.

næus and *Planorbis*, one of which is probably the parent of those eggs. And here is one which lays no eggs, but brings forth its young alive: it is the *Paludina vivipara* (Fig. 15), of which we learned some interesting details last month.

Fig. 15.—Paludina Vivipara.

Scattered over the surface of the net and dead leaves are little dabs of dirty-looking jelly—some of them, instead of the dirty hue, are almost blood-red. Experience makes me aware that these dirty dabs are certainly Polypes—the *Hydra fusca* of systematists. I can't tell how it is I know them, nor how you may know them again. The power of recognition must be acquired by familiarity; and it is because men can't *begin* with familiarity, and can't recognize these Polypes without it, that so few persons really ever see them. But the familiarity may be acquired by a very simple method. Make it a rule to pop every unknown object into your wide-mouthed phial. In the water it will probably at once reveal its nature: if it be a Polype, it will expand its tentacles; if not, you can identify it at leisure on reaching home by the aid of pictures and descriptions. See, as I drop one of these into the water, it at once assumes the well-known shape of the Polype. And now we will see what these blood-red dabs may be; in spite of their unusual color, I can not help suspecting them to be Polypes also. Give me the camel-hair brush. Gently the dab is removed, and transferred to the phial. Shade of Trembley! it *is* a Polype!* Is it possible that this discovery leaves you imperturbable, even

---

* TREMBLEY, in his admirable work, *Mémoires pour servir à l'histoire d'une genre de Polypes d'eau douce*, 1744, furnished science with the fullest and most accurate account of fresh-water Polypes; but it is a mistake to suppose that he was the original discoverer of this genus: old LEUWENHOEK had been before him.

when I assure you it is of a species hitherto undescribed in text-books? Now don't be provokingly indifferent! rouse yourself to a little enthusiasm, and prove that you have something of the naturalist in you by delighting in the detection of a new species. "You didn't know that it was new?" *That* explains your calmness. There must be a basis of knowledge before wonder can be felt—wonder being, as Bacon says, "broken knowledge." Learn, then, that hitherto only three species of fresh-water Polypes have been described: *Hydra viridis*, *Hydra fusca*, and *Hydra grisea*. We have now a fourth to swell the list; we will christen it *Hydra rubra*, and be as modest in our glory as we can. If any one puts it to us whether we seriously attach importance to such trivialities as specific distinctions resting solely upon color or size, we can look profound, you know, and repudiate the charge. But this is a public and official attitude. In private we can despise the distinctions established by others, but keep a corner of favoritism for our own.*

I remember once showing a bottle containing Polypes to a philosopher, who beheld them with great calmness. They appeared to him as insignifi-

---

* The editors of the *Annals of Natural History* append a note to the account I sent them of this new Polype, from which it appears that Dr. Gray found this very species, and apparently in the same spot, nearly thirty years ago. But the latest work of authority, Van der Hoven's *Handbook of Zoology*, only enumerates the three species.

cant as so many stems of duckweed; and, lest you should be equally indifferent, I will at once inform you that these creatures will interest you as much as any that can be found in ponds, if you take the trouble of studying them. They can be cut into many pieces, and each piece will grow into a perfect Polype; they may be pricked or irritated, and the irritated spot will bud a young Polype, as a plant buds; they may be turned inside out, and their skin will become a stomach, their stomach a skin. They have acute sensibility to light (toward which they always move), and to the slightest touch; yet not a trace of a nervous tissue is to be found in them. They have powers of motion and locomotion, yet their muscles are simply a network of large contractile cells. If the water in which they are kept be not very pure, they will be found infested with parasites; and quite recently I have noticed an animal or vegetal parasite—I know not which—forming an elegant sort of fringe to the tentacles; clusters of skittle-shaped bodies, too entirely transparent for any structure whatever to be made out, in active agitation, like leaves fluttering on a twig. Some day or other we may have occasion to treat of the Polypes in detail, and to narrate the amusing story of their discovery; but what has already been said will serve to sharpen your attention, and awaken some curiosity in them.

Again and again the net sweeps among the weed or dredges the bottom of the pond, bringing up mud, stones, sticks, with a fish, worms, mollusks, and tritons. The fish we must secure, for it is a stickleback—a pretty and interesting inhabitant of an aquarium, on account of its nest-building propensities. We are surprised at a fish building a nest and caring for its young like the tenderest of birds (and there are two other fishes, the Goramy and the Hassar, which have this instinct); but why not a fish as well as a bird? The catfish swims about in company with her young, like a proud hen with her chickens, and the sunfish hovers for weeks over her eggs, protecting them against danger.

The wind is so piercing, and my fingers are so benumbed, I can scarcely hold the brush. Moreover, continual stooping over the net makes the muscles ache unpleasantly, and suggests that each cast shall be the final one. But somehow I have made this resolution and broken it twenty times: either the cast has been unsuccessful, and one is provoked to try again, or it is so successful that, as *l'appétit vient en mangeant*, one is seduced again. Very unintelligible this would be to the passersby, who generally cast contemptuous glances at us when they find we are not fishing, but are only removing nothings into a glass jar. One day an Irish laborer stopped and asked me if I were fishing for salmon. I quietly answered "Yes." He drew near. I continued turning over the weed, occasionally dropping an invisible thing into the water. At last a large yellow-bellied Triton was dropped in. He begged to see it; and, seeing at the same time how alive the water was with tiny animals, became curious, and asked many questions. I went on with my work; his interest and curiosity increased; his questions multiplied; he volunteered assistance; and remained beside me till I prepared to go away, when he said seriously, "Och! then, and it's a fine thing to be able to name all God's creatures." Contempt had given place to reverence; and so it would be with others, could they check the first rising of scorn at what they do not understand, and patiently learn what even a roadside pond has of Nature's wonders.

CHAPTER III.

A garden Wall, and its Traces of past Life.—Not a Breath perishes.—A Bit of dry Moss and its Inhabitants.—The "Wheelbearers."—Resuscitation of Rotifers: drowned into Life.—Current Belief that Animals can be revived after complete Desiccation.—Experiments contradicting the Belief.—Spallanzani's Testimony.—Value of Biology as a Means of Culture.—Classification of Animals: the five great Types.—Criticism of Cuvier's Arrangement.

PLEASANT, both to eye and mind, is an old garden wall, dark with age, gray with lichens, green with mosses of beautiful hues and fairy elegance of form; a wall shutting in some sequestered home, far from "the din of murmurous cities vast;" a home where, as we fondly, foolishly think, Life must needs throb placidly, and all its tragedies and pettinesses be unknown. As we pass alongside this wall, the sight of the overhanging branches suggests an image of some charming nook; or our thoughts wander about the wall itself, calling up the years during which it has been warmed by the sun, chilled by the night airs and the dews, and dashed against by the wild winds of March, all of which have made it quite another wall from what it was when the trowel first settled its bricks. The old wall has a past, a life, a story; as Wordsworth finely says of the mountain, it is "familiar with forgotten years." Not only are there obvious traces of age in the crumbling mortar and the battered brick, but there are traces, not obvious except to the inner eye, left by every ray of light, every raindrop, every gust. Nothing perishes. In the wondrous metamorphosis momently going on everywhere in the universe, there is *change*, but no *loss*.

Lest you should imagine this to be poetry, and not science, I will touch on the evidence that every beam of light, or every breath of air which falls upon an object, permanently affects it. In photography we see the effect of light very strikingly exhibited; but perhaps you will object that this proves nothing more than that light acts upon an iodized surface. Yet, in truth, light acts upon, and more or less alters the structure of every object on which it falls. Nor is this all. If a wafer be laid on a surface of polished metal, which is then breathed upon, and if, when the moisture of the breath has evaporated, the wafer be shaken off, we shall find that the whole polished surface is not as it was before, although our senses can detect no difference; for if we breathe again upon it, the surface will be moist every where except on the spot previously sheltered by the wafer, which will now appear as a spectral image on the surface. Again and again we breathe, and the moisture evaporates, but still the spectral wafer reappears. This experiment succeeds after a lapse of many months if the metal be carefully put aside where its surface can not be disturbed. If a sheet of paper on which a key has been laid be exposed for some minutes to the sunshine, and then instantaneously viewed in the dark, the key being removed, a fading spectre of the key will be visible. Let this paper be put aside for many months where nothing can disturb it, and then in darkness be laid on a plate of hot metal, the spectre of the key will again appear. In the case of bodies more highly phosphorescent than paper, the spectres of many different objects which may have been laid on in succession will, on warming, emerge in their proper order.*

This is equally true of our bodies and our minds. We are involved in the universal metamorphosis. Nothing leaves us wholly as it found us. Every man we meet, every book we read, every picture or landscape we see, every word or tone we hear, mingles with our being and modifies it. There are cases on record of ignorant women, in states of insanity, uttering Greek and Hebrew phrases, which in past years they had heard their masters utter, without, of course, comprehending them. These tones had long been forgotten; the traces were so faint that under ordinary conditions they were invisible; but the traces were there, and in the intense light of cerebral excitement they started into prominence, just as the spectral image of the key started into sight on the application of heat. It is

* DRAPER: *Human Physiology*, p. 288.

thus with all the influences to which we are subjected.

If a garden wall can lead our vagabond thoughts into such speculations as these, surely it may also furnish us with matter for our Studies in Animal Life. Those patches of moss must be colonies. Suppose we examine them. I pull away a small bit, which is so dry that the dust crumbles at a touch; this may be wrapped in a piece of paper—dirt and all—and carried home. Get the microscope ready, and now attend.

I moisten a fragment of this moss with distilled water. Any water will do as well, but the use of distilled water prevents your supposing that the animals you are about to watch were brought in it, and were not already in the moss. I now squeeze the bit between my fingers, and a drop of the contained water—somewhat turbid with dirt—falls on the glass slide, which we may now put on the microscope stage. A rapid survey assures us that there is no animal visible. The moss is squeezed again, and this time little yellowish bodies of an irregular oval are noticeable among the particles of dust and moss. Watch one of these, and presently you will observe a slow bulging at one end, and then a bulging at the other end. The oval has elongated itself into a form not unlike that of a fat caterpillar, except that there is a tapering at one end. Now a forked tail is visible; this fixes on to the glass, while the body swings to and fro.

Now the head is drawn in—as if it were swallowed—and suddenly in its place are unfolded two broad membranes, having each a circle of waving *cilia*. The lifeless oval has become a living animal! You have assisted at a resuscitation, not from death by drowning, but by drying: the animal has been drowned into life! The unfolded membranes, with their cilia, have so much the appearance of wheels that the name of "Wheel-bearer" (*Rotifera*) or "Wheel Animalcule" has been given to the animal.

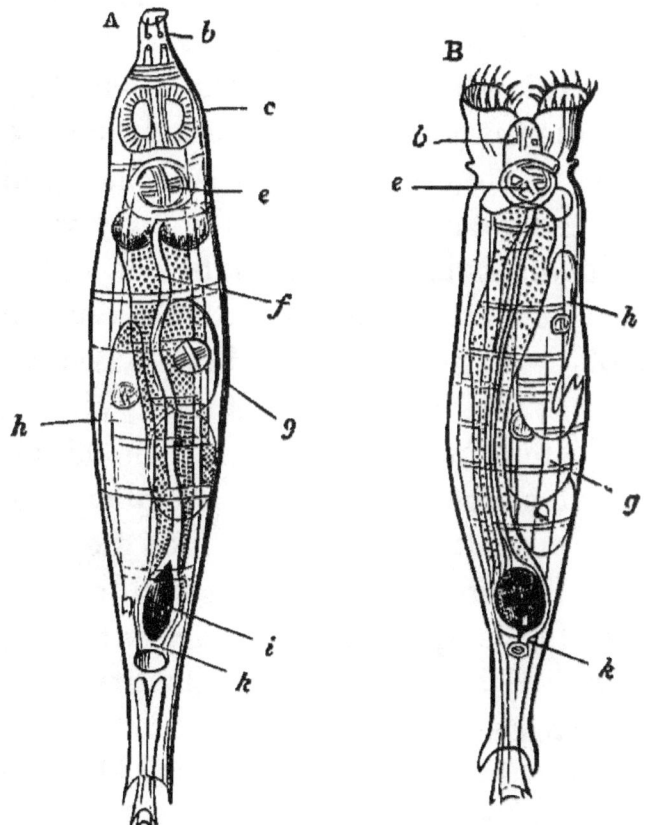

Fig. 16.—ROTIFER VULGARIS.

A, with the wheels drawn in (at *c*). B, with the wheels expanded; *b*, eye spots; *e*, jaw and teeth; *f*, alimentary canal; *g*, embryo; *h*, embryo further developed; *i*, water-vascular system; *k*, vent.

The Rotifera (also, and more correctly, called *Rotatoria*) form an interesting study. Let us glance at their organization:

There are many different kinds of Rotifers, varying very materially in size and shape, the males, as was stated in the last chapter, being more imperfectly organized than the females. They may be seen either swimming rapidly through the water by means of the vibratile cilia called "wheels," because the optical effect is very much that of a toothed wheel, or crawling along the side of the glass, fastening to it by the head, and then curving the body till the tail is brought up to the spot, which is then fastened on by the tail, and the head is set free. They may also be seen fastened to a weed, or the glass, by the tail, the body waving to and fro, or thrusting itself straight out, and setting the wheels in active motion. In this attitude the aspect of the jaws is very striking. Leuwenhoek mistook it for the pulsation of a heart, which its incessant rhythm much resembles. The tail and the upper part of the body have a singular power of being drawn out or drawn in, like the tube of a telescope. There is sometimes a shell or carapace, but often the body is covered only with a smooth firm skin, which, however, presents decided indications of being segmented.

The first person who described these Rotifers was the excellent old Leuwenhoek,[*] and his animals

---

[*] LEUWENHOEK: *Select Works*, ii., p. 210. His figures, however, are very incorrect.

were got from the gutter of a house-top. Since then they have been minutely studied, and have been shown to be, not Infusoria, as Ehrenberg imagined, but Crustacea.* Your attention is requested to the one point which has most contributed to the celebrity of these creatures—their power of resuscitation. Leuwenhoek described—what you have just witnessed, namely—the slow resuscitation of the animal (which seemed as dry as dust, and might have been blown about like any particle of dust) directly a little moisture was brought to it. Spallanzani startled the world with the announcement that this process of drying and moistening—of killing and reviving—could be repeated fifteen times in succession; so that the Rotifer, whose natural term of life is about eighteen days, might, it was said, be dried and kept for years, and at any time revived by moisture. That which seems now no better than a grain of dust will suddenly awaken to the energetic life of a complex organism, and may again be made as dust by the evaporation of the water.

This is very marvelous; so marvelous that a mind trained in the cultivated caution of science will demand the evidence on which it is based. Two months ago I should have dismissed the doubt with the assurance that the evidence was ample and

---

* See LEYDIG: *Ueber den Bau und die systematische Stellung der Räderthiere*, in SIEBOLD und KÖLLIKER'S *Zeitschrift*, vi., and *Ueber Hydatina Senta*, in MÜLLER'S *Archiv*, 1857.

rigorous, and the fact indisputable; for not only had the fact been confirmed by the united experience of several investigators, it had stood the test of very severe experiment. Thus, in 1842, M. Doyère published experiments which seemed to place it beyond skepticism. Under the air-pump he set some moss, together with vessels containing sulphuric acid, which would absorb every trace of moisture. After leaving the moss thus for a week, he removed it into an oven, the temperature of which was raised to 300° Fahrenheit. Yet even this treatment did not prevent the animals from resuscitating when water was added.

In presence of testimony like this, doubt will seem next to impossible. Nevertheless, my own experiments leave me no choice but to doubt. Not having witnessed M. Doyère's experiment, I am not prepared to say wherein its fallacy lies; but that there *is* a fallacy seems to me capable of decisive proof. In M. Pouchet's recent work* I first read a distinct denial of the pretended resuscitation of the Rotifers; this denial was the more startling to me, because I had myself often witnessed the reawakening of these dried animals. Nevertheless, whenever a doubt is fairly started, we have not done justice to it until we have brought it to the test of experiment; accordingly, I tested this, and quickly came upon what seems to me the source of the gen-

---

* POUCHET: *Hétérogénie, ou Traité de la Génération Spontanée,* 1859, p. 453.

eral misconception. Day after day experiments were repeated, varied, and controlled, and with results so unvarying that hesitation vanished; and as some of these experiments are of extreme simplicity, you may verify what I say with little trouble. Squeeze a drop from the moss, taking care that there is scarcely any dirt in it; and, having ascertained that it contains Rotifers or Tardigrades,* alive and moving, place the glass slide under a bell-glass, to shield it from currents of air, and there allow the water to evaporate slowly, but completely, by means of chloride of calcium or sulphuric acid placed under the bell-glass; or, what is still simpler, place a slide with the live animals on the mantelpiece when a fire is burning in the grate. If on the day following you examine this perfectly dry glass, you will see the contracted bodies of the Rotifers, presenting the aspect of yellowish oval bodies; but attempt to resuscitate them by the addition of a little fresh water, and you will find that they do not revive, as they revived when dried in the moss; they sometimes swell a little, and elongate themselves, and you imagine this is a commencement of resuscitation; but continue watching for two or three days, and you will find it goes no further.

* The *Tardigrade*, or microscopic *Sloth*, belongs to the order of Arachnida, and is occasionally found in moss, stagnant ponds, etc. I have only met with four specimens in all my investigations, and they were all found in moss. SPALLANZANI described and figured it (very badly), and M. DOYÈRE has given a fuller description in the *Annales des Sciences*, 2d series, vols. xiv., xvii., and xviii.

Never do these oval bodies become active crawling Rotifers; never do they expand their wheels, and set the œsophagus at work. No; the Rotifer once *dried* is dead, and dead forever.

But if, like a cautious experimenter, you vary and control the experiment, and beside the glass slide place a watch-glass containing Rotifers with dirt or moss, you will find that the addition of water to the contents of the watch-glass will often (not always) revive the animals. What you can not effect on a glass slide without dirt, or with very little, you easily effect in a watch-glass with dirt or moss; and if you give due attention you will find that in each case the result depends upon the quantity of the dirt. And this leads to a clear understanding of the whole mystery; this reconciles the conflicting statements. The reason why Rotifers ever revive is because they have not been *dried*—they have not lost by evaporation that small quantity of water which forms *an integral constituent of their tissues;* and it is the presence of dirt or moss which prevents this complete evaporation. No one, I suppose, believes that the Rotifer actually revives after once being dead. If it has a power of remaining in a state of suspended animation, like that of a frozen frog, it can do so only on the condition that its *organism* is not destroyed; and destroyed it would be if the water were removed from its tissues; for, strange as it may seem, water is not an *accessory*, but a *constituent element* of every tissue; and this can not be replaced *mechanically*—it can only be replaced by *vital processes*. Every one who has made microscopic preparations must be aware that when once a tissue is desiccated, it is spoiled; it will not recover its form and properties on the application of water, because the water was not originally worked into the web by a mere process of imbibition—like water in a sponge—but by a molecular process of assimilation, like albumen in a muscle. Therefore I say that desiccation is necessarily death, and the Rotifer which revives can not have been desiccated. This being granted, we have only to ask, What prevents the Rotifer from becoming completely dried? Experiment shows that it is the presence of dirt or moss which does this. The whole marvel of the Rotifer's resuscitation, therefore, amounts to this: that if the water in which it lives be evaporated, the animal passes into a state of suspended animation, and remains so as long as its *own water* is protected from evaporation.

I am aware that this is not easily to be reconciled with M. Doyère's experiment, since the application of a temperature so high as 300° Fahrenheit (nearly a hundred degrees above boiling water) must, one would imagine, have completely desiccated the animals, in spite of any amount of protecting dirt. It is possible that M. Doyère may have mistaken that previously-noticed swelling up of the bodies, on the application of water, for a return to vital activity. If not, I am at a loss to explain the contradiction;

for certainly in my experience a much more moderate desiccation—namely, that obtained by simple evaporation over a mantelpiece or under a large bell-glass—always destroyed the animals if little or no dirt were present.

The subject has recently been brought before the French Academy of Sciences by M. Davaine, whose experiments[*] lead him to the conclusion that those Rotifers which habitually live in ponds will not revive after desiccation, whereas those which live in moss always do so. I believe the explanation to be this: the Rotifers living in ponds are dried without any protecting dirt or moss, and that is the reason they do not revive.

After having satisfied myself on this point, I did what perhaps would have saved me some trouble if thought of before. I took down Spallanzani, and read his account of his celebrated experiments. To my surprise and satisfaction, it appeared that he had accurately observed the same facts, but curiously missed their real significance. Nothing can be plainer than the following passage: "But there is one condition indispensable to the resurrection of wheel-animals: it is absolutely necessary that there should be a certain quantity of sand; without it they will not revive. One day I had two wheel-animals traversing a drop of water about to evaporate which contained very little sand. Three quarters of an hour after evaporation they were dry and

[*] DAVAINE in *Annales des Sciences Naturelles*, 1858, x., p. 385.

motionless. I moistened them with water to revive them, but in vain, notwithstanding that they were immersed in water many hours. Their members swelled to thrice the original size, but they remained motionless. To ascertain whether the fact was accidental, I spread a portion of sand, containing animals, on a glass slide, and waited until it became dry in order to wet it anew. The sand was carelessly scattered on the glass, so as to be a thin covering on some parts, and on others in a very small quantity: here the animals did not revive; but all that were in those parts with abundance of sand revived."* He further says that if sand be spread out in considerable quantities in some places, much less in others, and very little in the rest, on moistening it the revived animals will be numerous in the first, less numerous in the second, and none at all in the third.

It is not a little remarkable that observations so precise as these should have for many years passed unregarded, and not led to the true explanation of the mystery. Perhaps an inherent love of the marvelous made men greedily accept the idea of resuscitation, and indisposed them to attempt an explanation of it. Spallanzani's own attempt is certainly not felicitous. He supposes that the dust prevents the lacerating influence of the air from irritating and injuring the animals. And this explanation is accepted by his translator.

* SPALLANZANI: *Tracts on the Natural History of Animals and Vegetables:* translated by Dalyell, ii., p. 129.

[Since the foregoing remarks were in type, M. Gavarret has published (*Annales des Sciences Naturelles*, 1859, xi., p. 315) the account of his experiments on Rotifers and Tardigrades, in which he found that after subjecting the *moss* to a desiccation the most complete according to our present means, the *animals* revived after twenty-four hours' immersion of the moss in water. This result seems flatly to contradict the result I arrived at, but only *seems* to contradict it, for in my experiments the *animals*, not the moss, were subjected to desiccation. Nevertheless, I confess that my confidence was shaken by experiments so precise, and performed by so distinguished an investigator, and I once more resumed the experiments, feeling persuaded that the detection of the fallacy, wherever it might be, would be well worth the trouble. The results of these controlling experiments are all I can find room for here: *Whenever the animals were completely separated from the dirt, they perished;* in two cases there was a very little dirt—a mere film, so to speak—in the watch-glass and glass cell, and this, slight as it was, sufficed to protect two out of eight, and three out of ten Rotifers, which revived on the second day; the others did not revive even on the third day after their immersion. In one instance, a thin covering-glass was placed over the water on the slide, and the evaporation of the water seemed complete, yet this glass cover sufficed to protect a Rotifer, which revived in three hours.

If we compare these results with those obtained by M. Davaine, we can scarcely avoid the conclusion that it is only when the desiccation of the Rotifers is prevented by the presence of a small quantity of moss or of dirt—between the particles of which they find shelter—that they revive on the application of water. And even in the severe experiments of M. Doyère and M. Gavarret, *some* of the animals must have been thus protected; and I call particular attention to the fact that, although some animals revived, others always perished. But if the organization of the Rotifer or Tardigrade is such that it can withstand desiccation—if it only needs the fresh application of moisture to restore its activity—all, or almost all the animals experimented on ought to revive; and the fact that only some revive leads us to suspect that these have not been desiccated—a suspicion which is warranted by direct experiments. I believe, then, that the discrepancy amounts to this: investigators who have desiccated the moss containing animals find some of the animals revive on the application of moisture, but those who desiccate the animals themselves will find no instances of revival.]

The time spent on these Rotifers will not have been misspent if it has taught us the necessity of caution in all experimental inquiries. Although experiment is valuable—nay, indispensable—as a means of interrogating Nature, it is constantly liable to mislead us into the idea that we have rightly interrogated and rightly interpreted the replies; and this danger arises from the complexity of the cases with which we are dealing, and our proneness to overlook or disregard some seemingly trifling condition—a trifle which may turn out of the utmost importance. The one reason why the study of science is valuable as a means of culture, over and above its own immediate objects, is that in it the mind learns to *submit* to realities instead of thrusting its figments in the place of realities—endeavors to ascertain accurately what the order of Nature *is*, and not what it ought to be or might be. The one reason why, of all sciences, Biology is preeminent as a means of culture, is, that, owing to the great complexity of all the cases it investigates, it familiarizes the mind with the necessity of attending to *all* the conditions, and it thus keeps the mind alert. It cultivates caution, which, considering the tendency there is in men to "anticipate Nature," is a mental tonic of inestimable worth. I am far from asserting that biologists are more accurate reasoners than other men; indeed, the mass of crude hypothesis which passes unchallenged by them is against such an idea. But, whether its advantage be used or neglected, the truth nevertheless is, that Biology, from the complexity of its problems, and the necessity of incessant verification of its details, offers greater advantages for culture than any other branch of science.

I have once or twice mentioned the words Mollusk and Crustacean, to which the reader unfamiliar with the language of Natural History will have attached but vague ideas; and although I wanted to explain these, and convey a distinct conception of the general facts of classification, it would have been too great an interruption. So I will here make an opportunity, and finish the chapter with an indication of the five types, or plans of structure, under one of which every animal is classed. Without being versed in science, you discern at once whether the book before you is mathematical, physical, chemical, botanical, or physiological. In like manner, without being versed in Natural History, you ought to know whether the animal before you belongs to the Vertebrata, Mollusca, Articulata, Radiata, or Protozoa.

A glance at the contents of our glass vases will yield us samples of each of these five divisions of the animal kingdom. We begin with this Triton (Fig. 17). It is a representative of the VERTEBRATE division or sub-kingdom. You have merely to remember that it possesses a backbone and an internal skeleton, and you will at once recognize the cardinal character which makes this Triton range under the same general head as men, elephants, whales, birds, reptiles, or fishes. All these, in spite of their manifold differences, have this one character in common—they are all backboned; they have all an internal skeleton; they are all formed according to one general type. In all vertebrate ani-

Fig. 17.—MALE TRITON, OR WATER-NEWT.

mals the skeleton is found to be identical in plan. Every bone in the body of a triton has its corresponding bone in the body of a man or of a mouse; and every bone preserves the same connection with other bones, no matter how unlike may be the various limbs in which we detect its presence. Thus, widely as the arm of a man differs from the fin of a whale, or the wing of a bird, or the wing of a bat, or the leg of a horse, the same number of bones, and the same connections of the bones, are found in each. A fin is one modified form of the typical limb; an arm is another; a wing another. That which is true of the limbs is also true of all the other organs; and it is on this ground that we speak of the vertebrate type. From fish to man one common plan of structure prevails, and the presence of a backbone is the index by which to recognize this plan.

The Triton has been wriggling grotesquely in our grasp while we have made him our text, and, now he is restored to his vase, plunges to the bottom with great satisfaction at his escape. This water-snail, crawling slowly up the side of the vase, and cleaning it of the green growth of microscopic plants, which he devours, shall be our representative of the second great division—the MOLLUSCA. I can not suggest any obvious character so distinctive as a backbone by which the word mollusk may at once call up an idea of the type which prevails in the group. It won't do to say "shellfish," because many mollusks have no shells, and many animals which have shells are not mollusks. The name was originally bestowed on account of the softness of the animals. But they are not softer than worms, and much less so than jellyfish. You may know that snails and slugs, oysters and cuttlefish, are mollusks; but if you want some one character by which the type may be remembered, you must fix on the imperfect symmetry of the mollusk's organs. I say *imperfect* symmetry, because it is an error, though a common one, to speak of the mollusk's body not being *bilateral*—that is to say, of its not being composed of two symmetrical halves. A vertebrate animal may be divided lengthwise, and each half will closely resemble the other; the backbone forms, as it were, an axis, on either side of which the organs are disposed; but the mollusk is said to have no such axis, no such symmetry. I admit the absence of an axis, but I deny the total absence of symmetry. Many of its organs are as symmetrical as those of a vertebrate animal—*i. e.*, the eyes, the feelers, the jaws—and the gills in Cuttlefish, Eolids, and Pteropods; while, on the other hand, several organs in the vertebrate animal are as *un*symmetrical as any of those in the mollusk—*i. e.*, the liver, spleen, pancreas, stomach, and intestines.* As regards bilateral structure, therefore, it is only a question of degree. The vertebrate animal is not entirely symmetrical, nor is the mollusk entirely unsymmetrical. But there is a characteristic disposition of the nervous system peculiar to mollusks: it neither forms an *axis* for the body, as it does in the Vertebrata and Articulata, nor a *centre*, as it does in the Radiata, but is altogether irregular and unsymmetrical. This will be intelligible from the following diagram of the nervous systems of a mollusk and an insect, with which that of a starfish may be compared (Fig. 18). Here you perceive how the nervous centres and the nerves which

---

* In some cases of monstrosity these organs are transposed, the liver being on the left, and the pancreas on the right side. It was in allusion to a case of this kind, then occupying the attention of Paris, that MOLIÈRE made his *Medecin malgré Lui* describe the heart as on the right side, the liver on the left; on the mistake being noticed, he replies, "*Oui, autrefois; mais nous avons changé tout cela.*"

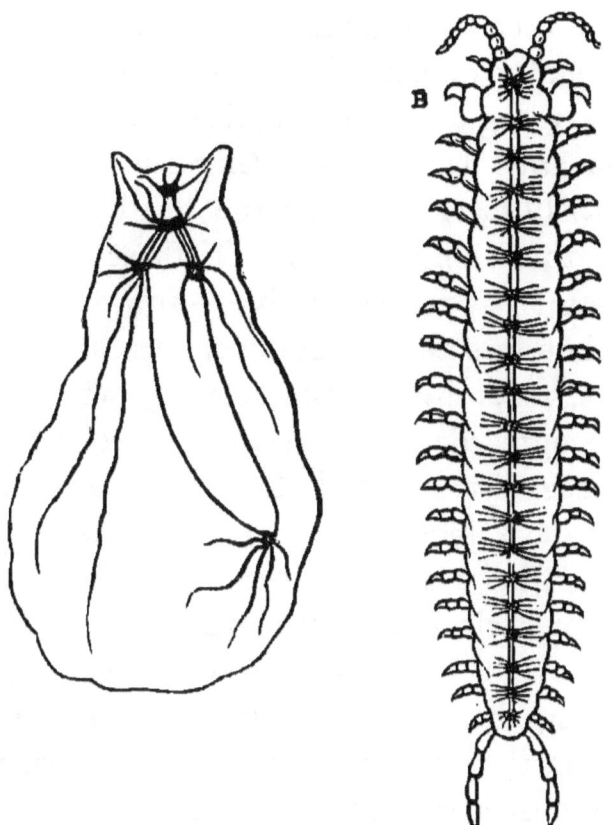

Fig. 18.—Nervous System of Sea-hare (A) and Centipede (B).

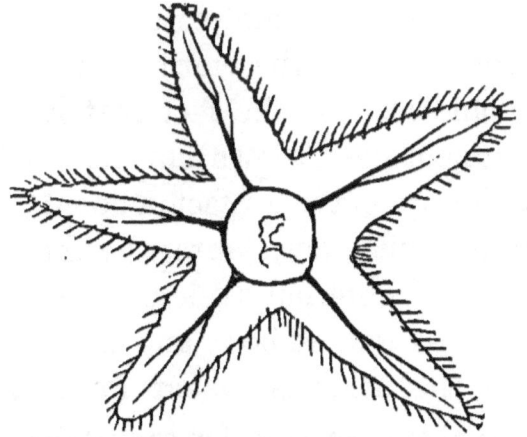

Fig. 19.—Nervous System of Starfish.

issue from them are irregularly disposed in the mollusks, and symmetrically in the insect.

But the recognition of a mollusk will be easier when you have learned to distinguish it from one of the ARTICULATA, forming the third great division—the third animal type. Of these, our vases present numerous representatives—prawns, beetles, water-spiders, insect-larvæ, entomostraca, and worms. There is a very obvious character by which these may be recognized: they have all bodies composed of numerous *segments*, and their limbs are *jointed*, and they have mostly an *external* skeleton from which their limbs are developed. Sometimes the segments of their bodies are numerous, as in the centipede, lobster, etc.; sometimes several segments are fused together, as in the crab; and sometimes, as in worms, they are indicated by slight markings or depressions of the skin, which give the appearance of little rings, and hence the worms have been named *Annelida*, or *Annulata*, or *Annulosa*. In these last-named cases the segmental nature of the type is detected in the fact that the worms grow, segment by segment; and also by the fact that in most of them each segment has its own nerves, heart, stomach, etc.—each segment is, in fact, a zöoid.*

Just as we recognize a vertebrate by the presence of a backbone and internal skeleton, we recognize an articulate by its jointed body and external skeleton. In both, the nervous system forms the axis

---

* The term zöoid was explained in our last chapter.

of the body. The Mollusk, on the contrary, has no skeleton, internal or external,* and its nervous system does not form an axis. As a rule, both vertebrates and articulates have limbs, although there are exceptions in serpents, fishes, and worms. The Mollusks have no limbs. Backboned, jointed, and non-jointed, therefore, are the three leading characteristics of the three types.

Let us now glance at the fourth division—the RADIATA, so called because of the disposition of the organs round a centre, which is the mouth. Our fresh-water vases afford us only *one* representative of this type—the *Hydra*, or fresh-water Polype, whose capture was recorded in the last chapter. Is it not strange that while *all* the Radiata are aquatic, not a single terrestrial representative having been discovered, only one should be found in fresh water? Think of the richness of the seas, with their hosts of Polypes, Actiniæ, Jellyfish, Starfishes, Sea-urchins, Sea-pens (*Pennatulæ*), Lily-stars (*Comatulæ*), and Sea-cucumbers (*Holothuriæ*), and then compare the poverty of rivers, lakes, and ponds, reduced to their single representative, the *Hydra*. The radiate structure may best be exhibited by the diagram of the nervous system of the Starfish† on page 79.

---

* In the cuttlefish there is the commencement of an internal skeleton in the cartilage-plates protecting the brain.

† It is right to add that there are serious doubts entertained respecting the claim of a starfish to the possession of a nervous sys-

Cuvier, to whom we owe this classification of the animal kingdom into four great divisions, would have been the first to recognize the chaotic condition in which he left this last division, and would have acquiesced in the separation of the PROTOZOA, which has since been made. This fifth division includes many of the microscopic animals known as *Infusoria*, and receives its name from the idea that these simplest of all animals represent, as it were, the beginnings of life.*

But Cuvier's arrangement is open to a more serious objection. The state of science in his day excused the imperfection of classing the Infusoria and parasites under the Radiata; but it was owing, I conceive, to an unphilosophical view of morphology that he placed the mollusks next to the Vertebrata, instead of placing the Articulata in that position. He was secretly determined by the desire to show that there are four very distinct types, or plans of structure, which can not by any transitions be brought under one law of development. Lamarck and Geoffroy St. Hilaire maintained the idea of unity of composition throughout the animal kingdom: in other words, that all the varieties of animal forms were produced by successive modifications; and several of the German naturalists maintained that the vertebrata in their embryonic stages passed

---

tem at all; but the radiate structure is represented in the diagram, as it also is, very clearly, in a Sea-anemone.

* Protozoa, from *proton*, first, and *zoon*, animal.

through forms which were permanent in the lower animals. This idea Cuvier always opposed. He held that the four types were altogether distinct; and by his arrangement of them, their distinctness certainly appears much greater than would be the case on another arrangement. But, without discussing this question here, it is enough to point out the fact of the enormous superiority in intelligence, in sociality, and in complexity of animal functions which insects and spiders exhibit when compared with the highest of the mollusks, to justify the removal of the mollusca, and the elevation of the articulata to the second place in the animal hierarchy. Nor is this all. If we divide animals into four groups, these four naturally dispose themselves into two larger groups: the first of these, comprising Vertebrata and Articulata, is characterized by a *nervous axis* and a *skeleton;* the second, comprising Mollusca and Radiata, is characterized by the absence of both nervous axis and skeleton. It is obvious that a bee much more closely resembles a bird than any mollusk resembles any vertebrate. If there are many and important differences between the vertebrate and articulate types, there are also many and important resemblances; if the nervous axis is *above* the viscera, and forms the dorsal line of the vertebrate, whereas it is *underneath* the viscera, and forms the ventral line in the articulate, it is, nevertheless, in both, the axis of the body, and in both it sends off nerves to supply symmetrical limbs; in both it has similar functions. And while the articulata thus approach in structure the vertebrate type, the mollusca are not only removed from that type by many diversities, but a number of them have such affinities with the Radiate type that it is only in quite recent days that the whole class of Polyzoa (or Bryozoa, as they are also called) has been removed from the Radiata, and ranged under the Mollusca.

To quit this topic, and recur once more to the five divisions, we have only the broad outlines of the picture in Vertebrata, Mollusca, Articulata, Radiata, and Protozoa; but this is a good beginning, and we can now proceed to the further subdivisions. Each of these five sub-kingdoms is divided into classes; these, again, into orders; these into families; these into genera; these into species; and these, finally, into varieties. Thus, suppose a dwarf terrier is presented to us with a request that we should indicate its various titles in the scheme of classification: we begin by calling it a vertebrate; we proceed to assign its class as the mammalian; its order is obviously that of the carnivora; its family is that of the fox, wolf, jackal, etc., named *Canidæ;* its genus is, of course, that of *Canis;* its species, terrier; its variety, dwarf terrier. Inasmuch as all these denominations are the expressions of scientific research, and not at all arbitrary or fanciful, they imply an immense amount of labor and sagacity in their establishment; and when we remember that naturalists have thus classed upward of half a million of distinct species, it becomes an interesting inquiry, What has been the guiding principle of this successful labor? on what basis is so large a superstructure raised? This question we shall answer in the next chapter.

CHAPTER IV.

An extinct Animal recognized by its Tooth: how came this to be possible?—The Task of Classification.—Artificial and natural Methods.—Linnæus, and his Baptism of the Animal Kingdom: his Scheme of Classification.—What is there underlying all true Classification?—The chief Groups.—What is a Species?—Restatement of the Question respecting the Fixity or Variability of Species.—The two Hypotheses.—Illustration drawn from the Romance Languages.—Caution to Disputants.

I WAS one day talking with Professor Owen in the Hunterian Museum, when a gentleman approached with a request to be informed respecting the nature of a curious fossil which had been dug up by one of his workmen. As he drew the fossil from a small bag, and was about to hand it for examination, Owen quietly remarked, "That is the third molar of the under jaw of an extinct species of rhinoceros." The astonishment of the gentleman at this precise and confident description of the fossil, before even it had quitted his hands, was doubtless very great. I know that mine was, until the reflection occurred that if some one, little acquainted with editions, had drawn a volume from his pocket, declaring he had found it in an old chest, any bibliophile would have been able to say at a glance, "That is an Elzevir;" or, "That is one of the Tauchnitz classics, stereotyped at Leipzig." Owen is as familiar with the aspect of the teeth of animals, living and extinct, as a student is with the aspect of editions. Yet, before that knowledge could have been acquired, before he could say thus confidently that the tooth belonged to an extinct species of rhinoceros, the united labors of thousands of diligent inquirers must have been directed to the classification of animals. How could he know that the rhinoceros was of that particular species rather than another? and what is meant by species? To trace the history of this confidence would be to tell the long story of zoological investigation; a story too long for narration here, though we may pause a while to consider its difficulties.

To make a classical catalogue of the books in the British Museum would be a gigantic task; but imagine what that task would be if all the title-pages and other external indications were destroyed! The first attempts would necessarily be of a rough approximative kind, merely endeavoring to make a sort of provisional order amid the chaos, after which succeeding labors might introduce better and better arrangements. The books might first be grouped according to size; but, having got them together, it would soon be discovered that size was no indication of their contents: quarto poems and duodecimo histories, octavo grammars and folio dictionaries, would immediately give warning that some other arrangement was needed. Nor would it be better to separate the books according to the languages in which they were written. The presence or absence of "illustrations" would furnish no better guide, while the bindings would soon be found to follow no rule. Indeed, one by one, all the external characters would prove unsatisfactory, and the laborers would finally have to decide upon some internal characters. Having read enough of each book to ascertain whether it was poetry or prose—and, if poetry, whether dramatic, epic, lyric, or satiric; and if prose, whether history, philosophy, theology, philology, science, fiction, or essay— a rough classification could be made; but even then there would be many difficulties, such as where to place a work on the philosophy of history—or the history of science—or theology under the guise of science—or essays on very different subjects, while some works would defy classification.

Gigantic as this labor would be, it would be trifling compared with the labor of classifying all the animals now living (not to mention extinct species), so that the place of any one might be securely and rapidly determined; yet the persistent zeal and sagacity of zoologists have done for the animal kingdom what has not yet been done for the library of the Museum, although the titles of the books are not absent. It has been done by patient *reading* of the contents—by anatomical investigation of the internal structure of animals. Except on a basis of comparative anatomy, there could have been no better a classification of animals than a classification of books according to size, language, binding, etc.

An unscientific Pliny might group animals according to their habitat; but when it was known that whales, though living in the water and swimming like fishes, were in reality constructed like air-breathing quadrupeds—when it was known that animals differing so widely as bees, birds, bats, and flying squirrels, or as otters, seals, and cuttlefish, lived together in the same element, it became obvious that such a principle of arrangement could lead to no practical result. Nor would it suffice to class animals according to their modes of feeding, since in all classes there are samples of each mode. Equally unsatisfactory would be external form—the seal and the whale resembling fishes, the worm resembling the eel, and the eel the serpent.

Two things were necessary: first, that the structure of various animals should be minutely studied and described—which is equivalent to reading the books to be classified; and, secondly, that some artificial method should be devised of so arranging the immense mass of details as to enable them to be remembered, and also to enable fresh discoveries readily to find a place in the system. We may be perfectly familiar with the contents of a book, yet wholly at a loss where to place it. If we have to catalogue Hegel's *Philosophy of History*, for example, it becomes a difficult question whether to place it under the rubric of philosophy, or under that of history. To decide this point, we must have some system of classification.

In the attempts to construct a system, naturalists are commonly said to have followed two methods, the artificial and the natural. The *artificial method* seizes some one prominent characteristic, and groups all the individuals together which agree in this one respect. In Botany the artificial method classes plants according to the organs of reproduction; but this has been found so very imperfect that it has been abandoned, and the *natural method* has been substituted, according to which the whole structure of the plant determines its place. If flying were taken as the artificial basis for the grouping of some animals, we should find insects and birds, bats and flying squirrels grouped together; but the natural method, taking into consideration not one character, but all the essential characters, finds that insects, birds, and bats differ profoundly in their organization: the insect has wings, but its wings are not formed like those of the bird, nor are those of the bird formed like those of the bat. The insect does not breathe by lungs, like the bird and the bat; it has no internal skeleton, like the bird and the bat; and the bird, although it has many points in common with the bat, does not, like it, suckle its young; and thus we may run over the characters of each organization, and find that the three animals belong to widely different groups.

It is to Linnæus that we are indebted for the most ingenious and comprehensive of the many schemes invented for the cataloguing of animal forms, and modern attempts at classification are only improvements on the plan he laid down. First we may notice his admirable invention of the double names. It had been the custom to designate plants and animals according to some name common to a large group, to which was added a description more or less characteristic. An idea may be formed of the necessity of a reform by conceiving what a laborious and uncertain task it would be if our friends spoke to us of having seen a dog in the garden, and on our asking what kind of a dog, instead of their saying "a terrier, a bull-terrier, or a Skye-terrier," they were to attempt a description of the dog. Something of this kind was the labor of understanding the nature of an animal from the vague description of it given by naturalists. Linnæus rebaptized the whole animal kingdom upon one intelligible principle. He continued to employ the name common to each group, such as that of *Felis* for the cats, which became the *generic* name; and in lieu of the *description* which was given of each different kind to indicate that it was a lion, a tiger, a leopard, or a domestic cat, he affixed a *specific* name: thus the animal bearing the description of a lion became *Felis leo;* the tiger, *Felis tigris;* the leopard, *Felis leopardus;* and our domestic friend, *Felis catus*. These double names, as Vogt remarks, are like the Christian- and sur-names by which we distinguish the various members of one family; and instead of speaking of Tomkinson with the flabby face and Tomkinson with the square forehead, we simply say John and William Tomkinson.

Linnæus did more than this. He not only fixed definite conceptions of species and genera, but introduced those of orders and classes. Cuvier added families to genera, and sub-kingdoms (*embranchements*) to classes. Thus a scheme was elaborated by which the whole animal kingdom was arranged in subordinate groups: the sub-kingdoms were divided into classes, the classes into orders, the orders into families, the families into genera, the genera into species, and the species into varieties. The guiding principle of anatomical resemblance determined each of these divisions. Those largest groups, which resemble each other only in having what is called the typical character in common, are brought together under the first head. Thus all the groups which agree in possessing a backbone and internal skeleton, although they differ widely in form, structure, and habitat, do nevertheless resemble each other more than they resemble the groups which have no backbone. This great division having been formed, it is seen to arrange itself in very obvious minor divisions or classes—the mammalia, birds, reptiles, and fishes. All mammals resemble each other more than they resemble birds; all reptiles resemble each other more than they resemble fishes (in spite of the superficial resemblance between serpents and eels or lampreys). Each class, again, falls into the minor groups of orders, and on the same principles—the monkeys being obviously distinguished from rodents, and the carnivora from the ruminating animals; and so of the rest. In each order there are generally families, and the families fall into genera, which differ from each other only in fewer and less important characters. The genera include groups which have still fewer differences, and are called species; and these, again, include groups which have only minute and unimportant differences of color, size, and the like, and are called sub-species, or varieties.

Whoever looks at the immensity of the animal kingdom, and observes how intelligibly and systematically it is arranged in these various divisions, will admit that, however imperfect, the scheme is a magnificent product of human ingenuity and labor. It is not an arbitrary arrangement, like the grouping of the stars in constellations; it expresses, though obscurely, the real order of Nature. All true classification should be to forms what laws are to phenomena; the one reducing varieties to systematic order, as the other reduces phenomena to their relation of sequence. Now if it be true that the classification expresses the real order of Nature, and not simply the order which we may find convenient, there will be something more than mere resemblance indicated in the various groups; or, rather let me say, this resemblance itself is the consequence of some community in the things compared, and will therefore be the mark of some deeper cause. What is this cause? Mr. Darwin holds that "propinquity of descent — the only known cause of the similarity of organic beings—is the bond, hidden as it is by various degrees of modification, which is partially revealed to us by our classifications"*—" that the characters which naturalists consider as showing true affinity between any two or more species are those which have been inherited from a common parent, and in so far all true classification is genealogical; that community of descent is the hidden bond which naturalists have been unconsciously seeking, and not some unknown plan of creation, or the enunciation of general propositions, and the mere putting together and separating objects more or less alike."†

Before proceeding to open the philosophical discussion which inevitably arises on the mention of Mr. Darwin's book, I will here set down the chief groups, according to Cuvier's classification, for the benefit of the tyro in natural history, who will easily remember them, and will find the knowledge constantly invoked.

There are four sub-kingdoms, or branches: 1. Vertebrata; 2. Mollusca; 3. Articulata; 4. Radiata.

The VERTEBRATA consist of four classes: Mammalia, Birds, Reptiles, and Fishes.

The MOLLUSCA consist of six classes: Cephalopoda (cuttlefish), Pteropoda, Gasteropoda (snails,

---

* DARWIN: *Origin of Species*, p. 414. † Ibid., p. 420.

etc.), Acephala (oysters, etc.), Brachiopoda, and Cirrhopoda (barnacles).—N.B. This last class is now removed from the Mollusks and placed among the Crustaceans.

The ARTICULATA are composed of four classes: Annelids (worms), Crustacea (lobsters, crabs, etc.), Arachnida (spiders), and Insecta.

The RADIATA embrace all the remaining forms; but this group has been so altered since Cuvier's time that I will not burden your memory just now with an enumeration of the details.

The reader is now in a condition to appreciate the general line of argument adopted in the discussion of Mr. Darwin's book, which is at present exciting very great attention, and which will, at any rate, aid in general culture by opening to many minds new tracts of thought. The benefit in this direction is, however, considerably lessened by the extreme vagueness which is commonly attached to the word "species," as well as by the great want of philosophic culture which impoverishes the majority of our naturalists. I have heard or read few arguments on this subject which have not impressed me with the sense that the disputants really attached no distinct ideas to many of the phrases they were uttering. Yet it is obvious that we must first settle what are the facts grouped together and indicated by the word "species," before we can carry on any discussion as to the origin of species. To be battling about the fixity or variability of species, without having rigorously settled *what* species is, can lead to no edifying result.

It is notorious that if you ask even a zoologist, *What* is a species? you will always find that he has only a very vague answer to give; and if his answer be precise, it will be the precision of error, and will vanish into contradictions directly it is examined. The consequence of this is, that even the ablest zoologists are constantly at variance as to specific characters, and often can not agree whether an animal shall be considered of a new species or only a variety. There could be no such disagreements if specific characters were definite—if we knew *what* species meant, once and for all. Ask a chemist, What is a salt? What an acid? and his reply will be definite and uniformly the same: what he says all chemists will repeat. Not so the zoologist. Sometimes he will class two animals as of different species, when they only differ in color, in size, or in the numbers of tentacles, etc.; at other times he will class animals as belonging to the same species, although they differ in size, color, shape, instincts, habits, etc. The dog, for example, is said to be one species with many varieties or races. But contrast the pug-dog with the greyhound, the spaniel with the mastiff, the bull-dog with the Newfoundland, the setter with the terrier, the sheep-dog with the pointer; note the striking differences in their structure and their instincts, and you will find that they differ as widely as some genera and as some species. If these varieties inhabited different countries—if the pug were peculiar to Australia and the mastiff to Spain—there is not a naturalist but would class them as of different species. The same remark applies to pigeons and ducks, oxen and sheep.

The reason of this uncertainty is that the *thing* species does not exist: the term expresses an *abstraction*, like virtue, or whiteness; not a definite concrete reality, which can be separated from other things, and always be found the same. Nature produces individuals; these individuals resemble each other in varying degrees; according to their resemblances we group them together as classes, orders, genera, and species; but these terms only express the *relations of resemblance*, they do not indicate the existence of such *things* as classes, orders, genera, or species.* There is a reality indicated by each term —that is to say, a real relation; but there is no objective existence of which we could say, This is variable, This is immutable. Precisely as there is a real relation indicated by the term goodness, but there is no goodness apart from the virtuous actions and feelings which we group together under this term. It is true that metaphysicians in past ages angrily debated respecting the immutability of virtue, and had no more suspicion of their absurdity than moderns have who debate respecting the fixity

---

* CUVIER says, in so many words, that classes, orders, and genera are abstractions, *et rien de pareil n'existe dans la nature;* but species is *not* an abstraction !—See *Lettres à Pfaff*, p. 179.

of species. Yet no sooner do we understand that species means a relation of resemblance between animals, than the question of the fixity or variability of species resolves itself into this: Can there be any *variation in the resemblances* of closely allied animals? A question which would never be asked.

No one has thought of raising the question of the fixity of varieties, yet it is as legitimate as that of the fixity of species; and we might also argue for the fixity of genera, orders, classes, the fixity of all these being implied in the very terms; since no sooner does any departure from the type present itself, than *by* that it is excluded from the category; no sooner does a white object become gray or yellow, than it is excluded from the class of white objects. Here, therefore, is a sense in which the phrase "fixity of species" is indisputable; but in this sense the phrase has never been disputed. When zoologists have maintained that species are variable, they have meant that *animal forms are variable;* and these variations, gradually accumulating, result at last in such differences as are called specific. Although some zoologists, and speculators who were not zoologists, have believed that the possibility of variation is so great that one species may actually be *transmuted* into another, *i. e.*, that an ass may be developed into a horse, yet most thinkers are now agreed that such violent changes are impossible, and that every new form becomes established only through the long and gradual accumulation of minute differences in divergent directions.

It is clear, from what has just been said, that the many angry discussions respecting the fixity of species, which, since the days of Lamarck, have disturbed the amity of zoologists and speculative philosophers, would have been considerably abbreviated had men distinctly appreciated the equivoque which rendered their arguments hazy. I am far from implying that the battle was purely a verbal one. I believe there was a real and important distinction in the doctrines of the two camps; but it seems to me that, had a clear understanding of the fact that species was an abstract term been uniformly present to their minds, they would have sooner come to an agreement. Instead of the confusing disputes as to whether one species could ever become another species, the question would have been, Are animal forms changeable? Can the descendants of animals become so *unlike their ancestors*, in certain peculiarities of structure or instinct, as to be classed by naturalists as a different species?

No sooner is the question thus disengaged from equivoque than its discussion becomes narrowed within well-marked limits. That animal forms *are* variable is disputed by no zoologist. The only question which remains is this: *To what extent* are animal forms variable? The answers given have been two: one school declaring that the extent of variability is limited to those trifling characteristics which mark the different varieties of each species; the other school declaring that the variability is indefinite, and that all animal forms may have arisen from successive modifications of a very few types, or even of one type.

Now I would call your attention to one point in this discussion which ought to be remembered when antagonists are growing angry and bitter over the subject; it is, that both these opinions are necessarily hypothetical—there can be nothing like positive proof adduced on either side. The utmost that either hypothesis can claim is that it is more consistent with general analogies, and better serves to bring our knowledge of various points into harmony. Neither of them can claim to be a truth which warrants dogmatic decision.

Of these two hypotheses, the first has the weight and majority of authoritative adherents. It declares that all the different kinds of bats, for example, were distinct and independent creations, each species being originally what we see it to be now, and what it will continue to be as long as it exists: lions, panthers, pumas, leopards, tigers, jaguars, ocelots, and domestic cats being so many *original stocks*, and not so many *divergent forms of one original stock*. The second hypothesis declares that all these kinds of cats represent divergencies of the original stock, precisely as the varieties of each kind represent the divergencies of each species. It is true that each species, when once formed, only admits of limited variations; any cause which should push the variation *beyond* certain limits would destroy the species, because by species is meant the group of animals contained *within* those limits. Let us suppose the original stock from which all these kinds of cats have sprung to have become modified into lions, leopards, and tigers—in other words, that the gradual accumulation of divergencies has resulted in the whole family of cats existing under these three forms. The lions will form a distinct species; this species varies, and in the course of long variation a new species, the puma, rises by the side of it. The leopards also vary, and let us suppose their variation at length assumes so marked a form—in the ocelot—that we class it as a new species. There is nothing in this hypothesis but what is strictly consonant with analogies; it is only extending to species what we know to be the fact with respect to varieties; and these varieties which we know to have been produced from one and the same species are often more widely separated from each other than the lion is from the puma, or the leopard from the ocelot. Mr. Darwin remarks that " at least a score of pigeons might be chosen, which, if shown to an ornithologist, and he were told that they were wild birds, would certainly, I think, be ranked by him as well-defined species. Moreover, I do not believe that any ornithologist would place the English carrier, the short-faced tumbler, the runt, the barb, the pouter, and fantail in the same genus, more especially as in each of these breeds several truly-inherited sub-breeds or species, as he might have called them, could be shown him."

The development of numerous specific forms, widely distinguished from each other, out of one common stock, is not a whit more improbable than the development of numerous distinct languages out of a common parent language, which modern philologists have proved to be indubitably the case. Indeed, there is a very remarkable analogy between philology and zoology in this respect: just as the comparative anatomist traces the existence of similar organs, and similar connections of these organs, throughout the various animals classed under one type, so does the comparative philologist detect the family likeness in the various languages scattered from China to the Basque Provinces, and from Cape Comorin across the Caucasus to Lapland—a likeness which assures him that the Teutonic, Celtic, Wendic, Italic, Hellenic, Iranic, and Indic languages are of common origin, and separated from the Arabian, Aramean, and Hebrew languages, which have another origin. Let us bring together a Frenchman, a Spaniard, an Italian, a Portuguese, a Wallachian, and a Rhætian, and we shall hear six very different languages spoken, the speakers severally unintelligible to each other, their languages differing so widely that one can not be regarded as the modification of the other; yet we know most positively that all these languages are offshoots from the Latin, which was once a living language, but which is now, so to speak, a fossil. The various species of cats do not differ more than these six languages differ, and yet the resemblances point in each case to a common origin. Max Muller, in his brilliant essay on *Comparative Mythology*,* has said,

"If we knew nothing of the existence of Latin—if all historical documents previous to the fifteenth century had been lost—if tradition, even, was silent as to the former existence of a Roman empire, a mere comparison of the six Roman dialects would enable us to say that at some time there must have been a language from which all these modern dialects derived their origin in common; for without this supposition it would be impossible to account for the facts exhibited by these dialects. Let us look at the auxiliary verb. We find:

	Italian.	Wallachian.	Rhætian.	Spanish.	Portuguese.	French.
I am	sono	sum sunt	sunt	soy	sou	suis.
Thou art	sei	es	eis	eres	es	es.
He is	e	é (este)	ei	es	he	est.
We are	siamo	suntemu	essen	somos	somos	sommes.
You are	siete	sunteti	esses	sois	sois	êtes (estes).
They are	sono	sunt	eân (sun)	son	são	sont.

It is clear, even from a short consideration of these forms, first, that all are but varieties of one common type; secondly, that it is impossible to consider any one of these six paradigms as the original from which the others had been borrowed. To this we may add, thirdly, that in none of the languages to which these verbal forms belong do we find the

* See *Oxford Essays*, 1856.

elements of which they could have been composed. If we find such forms as *j'ai aimé*, we can explain them by a mere reference to the radical means which French has still at its command, and the same may be said even of compounds like *j'aimerai*, i. e., *je-aimer-ai*, I have to love, I shall love. But a change from *je suis* to *tu es* is inexplicable by the light of French grammar. These forms could not have grown, so to speak, on French soil, but must have been handed down as relics from a former period—must have existed in some language antecedent to any of the Roman dialects. Now, fortunately, in this case, we are not left to a mere inference, but as we possess the Latin verb, we can prove how, by phonetic corruption and by mistaken analogies, every one of the six paradigms is but a national metamorphosis of the Latin original.

"Let us now look at another set of paradigms:

	Sanscrit.	Lithu-anian.	Zend.	Doric.	Old Slavonic.	Latin.	Gothic.	Armen.
I am	ásmi	esmi	ahmi	ἐμμι	yesmŭ	sum	im	em.
Thou art	ási	essi	ahi	ἐσσί	yesi	es	is	es.
He is	ásti	esti	asti	ἐστί	yestŏ	est	ist	ê.
We (two) are	'avás	esva	....	....	yesva	...	siju	...
You (two) are	'sthás	esta	stho?	ἐστόν	yesta	...	sijuts	...
They (two) are	'stás	(esti)	sto?	ἐστόν	yesta	...	....	...
We are	'smás	esmi	hmahi	ἐσμές	yesmŏ	sumus	sijum	emq.
You are	'sthá	este	stha	ἐστέ	yeste	estis	sijup	êq.
They are	sánti	(esti)	hĕnti	ἐντί	somtŭ	sunt	sind	en.

"From a careful consideration of these forms, we ought to draw exactly the same conclusions; firstly, that all are but varieties of one common type; secondly, that it is impossible to consider any of them as the original from which the others have been borrowed; and, thirdly, that here again none of the languages in which these verbal forms occur possess the elements of which they are composed."

All these languages resemble each other so closely that they point to some more ancient language which was to them what Latin was to the six Roman languages; and in the same way we are justified in supposing that all the classes of the vertebrate animals point to the existence of some elder type, now extinct, from which they were all developed.

I have thus stated what are the two hypotheses on this question. There is only one more preliminary which it is needful to notice here, and that is, to caution the reader against the tendency, unhappily too common, of supposing that an adversary holds opinions which are transparently absurd. When we hear a hypothesis which is either novel or unacceptable to us, we are apt to draw some very ridiculous conclusion from it, and to assume that this conclusion is seriously held by its upholders. Thus the zoologists who maintain the variability of species are sometimes asked if they believe a goose was developed out of an oyster, or a rhinoceros from a mouse? the questioner apparently having no misgiving as to the candor of his ridicule. There are three modes of combating a doctrine. The first is to point out its strongest positions, and then show them to be erroneous or incomplete; but this plan is generally difficult, and sometimes impossible; it is not, therefore, much in vogue. The second is to render the doctrine ridiculous by pretending that it includes certain extravagant propositions of which it is entirely innocent. The third is to render the doctrine odious by forcing on it certain conclusions which it would repudiate, but which are declared to be " the inevitable consequences" of such a doctrine. Now it is undoubtedly true that men frequently maintain very absurd opinions; but it is neither candid nor wise to assume that men who otherwise are certainly not fools, hold opinions the absurdity of which is transparent.

Let us not, therefore, tax the followers of Lamarck, Geoffroy St. Hilaire, or Mr. Darwin with absurdities they have not advocated, but rather endeavor to see what solid argument they have for the basis of their hypothesis.

CHAPTER V.

Talking in Beetles.—Identity of Egyptian Animals with those now existing: Does this prove Fixity of Species?—Examination of the celebrated Argument of Species not having altered in four thousand Years.—Impossibility of distinguishing Species from Varieties.—The Affinities of Animals.—New Facts proving the Fertility of Hybrids.—The Hare and the Rabbit contrasted.—Doubts respecting the Development Hypothesis.—On Hypothesis in Natural History.—Pliny, and his Notion on the Formation of Pearls.—Are Pearls owing to a Disease of the Oyster?—Formation of the Shell; Origin of Pearls.—How the Chinese manufacture Pearls.

A WITTY friend of mine expressed her sense of the remoteness of the ancient Egyptians, and her difficulty in sympathizing with them, by declaring that "*they talked in beetles, you know.*" She referred, of course, to the hieroglyphics in which that curious people now speak to us from ancient tombs. Whether these swarthy sages were eloquent and wise, or obscure and otherwise, in their beetle-speech, it is certain that entomologists of our day recognize their beetles as belonging to the same species that are now gathered into collections. Such as the Egyptians knew them, such we know them now. Nay, the sacred cats found in those ancient tombs are cats of the same kind as our own familiar mousers; they purred before Pharaoh as they purr on our hearth-rugs; and the descendants of the very dogs which irreligiously worried those cats are to this day worrying the descendants of those sacred cats. The grains of wheat which the *savans* found in the tombs were planted in the soil of France, and grew into waving corn in no respect distinguishable from the corn grown from the grain of the previous year.

Have these familiar facts any important significance? Are we entitled to draw any conclusion from the testimony of paintings and sculptures, at least four thousand years old, which show that several of our well-known species of animals, and several of the well-marked races of men, existed then, and have not changed since then? Nimrod hunted with dogs and horses, which would be claimed as ancestors by the dogs and horses at Melton Mowbray. The negroes who attended Semiramis and Rhamses are in every respect similar to the negroes now toiling amid the sugar-canes of Alabama. If, during four thousand years, species and races have not changed, why should we suppose that they ever will change? Why should we not take our stand on that testimony, and assert that species are unchangeable?

Such has been the argument of Cuvier and his followers; an argument on which they have laid great stress, and which they have further strengthened by a challenge to adversaries to produce one single case where a *transmutation* of species has taken place: "Here we show you evidence that species has persisted unaltered during four thousand years, and you can not show us a single case of species having changed—you can not show us one case of a wolf becoming a dog, an ass becoming a horse, a hare becoming a rabbit. Yet you must admit that if there were any inherent tendency to change, four thousand years is a long enough period for that tendency to display itself in; and we ought to see a very marked difference between the species which lived under Semiramis and those which are living under Victoria. Instead of this, we see that there has been no change: the dog has remained a dog, and the horse has remained a horse; every species retains its well-marked characters."

No one will say that I have not done justice to this argument. I have stated it as clearly and forcibly as possible, not with any design to captivate your assent, but to make the answer complete. This argument is the *cheval de bataille* of the Cuvier school; but, like many other argumentative warhorses, it proves, on close inspection, to be spavined and broken-winded. The first criticism we must pass on it is that it implies the existence of species as a *thing* which can be spoken of as fixed or variable; whereas, as we saw in the last chapter, species is an *abstraction*, like whiteness or strength. No one supposes that there exists any whiteness apart from white things, or strength apart from strong things;

yet the naturalists who maintain the fixity of species constantly talk as if species existed independently of the individual animals. Instead of saying that by the word species is indicated a certain group of characters, and that whenever we meet with this group we say, here is an animal of the same species, they explicitly declare, or tacitly imply, that although an individual dog may vary, there is something above all individuals—the species—and *that* can not vary. As it is possible some readers may protest that no respectable authority in modern times ever held the opinion here imputed to a school, I will quote the very explicit language of one of Cuvier's disciples—the last editor of Buffon—who, no later than 1856, could declare that "species are the primitive forms of Nature. Individuals are nothing but the representatives—the copies of these forms: *Les espèces sont les formes primitives de la Nature. Les individus n'en sont que des représentations, des copies.*"* According to this very explicit but very extravagant statement, an individual dog is nothing but a copy of the primitive form—the typical dog—the *idea* of a dog, as Plato would say; and, of course, if this be true, it matters little how widely individual dogs may vary, the type, or species, of which it is the representative, remains unaltered. Indeed, it is on this ground that many physiologists explain the fact of hereditary transmission: the individual may vary, it is said, but

* FLOURENS: *Cours de Physiologie Comparée*, 1856, p. 9.

the species is preserved; and if a dog without its fore paws has offspring, every one of which possesses the fore paws, the reason is, that *l'idée de l'espèce se reproduit dans le fruit, et lui donne des organes qui manquaient au père ou à la mère.** It is not easy to understand how the *idea* of a species can reproduce itself, and give the offspring of a dog the *organs* which were wanting in the parents; but to those who believe that species exist independently of individuals, and form the only real existences, the conception may be easier.

I have too much respect for the reader to drag him through a refutation of such philosophy as this; the statement of the opinion is enough. And yet, unless some such opinion be maintained, the doctrine of fixity of species is without a basis; for if it be said that the group of characters which constitute the dog are incapable of change, and in this sense species are fixed, we have to ask what evidence there can be for such an assertion? since it is notorious that individual dogs *do* show a change in some of the characters of the group. We shall be referred to the Egyptian tombs for evidence. M. Flourens assures us that not only are these tombs evidence that species have not changed in four thousand years, but that *no* species has changed —*aucune espèce n'a changé*—which is surely stepping a long way beyond the precincts of the tombs!

It may be paradoxical, but it is strictly true, that

* BURDACH: *Physiologie*, li., 245.

the fact of particular species having remained unaltered during four thousand years does not add the slightest weight to the evidence in favor of the fixity of species. "What!" some may exclaim, "do you pretend that four thousand years is not a period long enough to prove the fixity of animal forms?" Yes; I affirm that four thousand, or forty thousand, prove no more than four. It is only by a fallacy that the opposite opinion could gain acceptance. You would not suppose that I had strengthened my case if, instead of contenting myself with stating reasons once, I repeated these same reasons during forty successive pages; you would remind me that this *iteration* was not *cumulation*, and that no force was given to my fortieth assertion which the first wanted. Why, then, do you ask me to accept the repetition of the same fact four thousand times over as an increase of evidence? It is a familiar fact that like produces like—that dogs resemble dogs, and do not resemble buffaloes; this fact is, of course, deepened in our conviction by the unvarying evidence we see around us, and is guaranteed by the philosophical axiom that like causes produce like effects; but when once such a conception is formed, it can gain no fresh strength from any particular instance. If we believe that crows are black, we do not hold that belief more firmly when we are shown that crows were black four thousand years ago. In like manner, if it is an admitted fact that individuals always reproduce individuals closely resembling themselves, it is not a whit more surprising that the dogs of Victoria should resemble the dogs of Semiramis, than that they should resemble their parents: the chain of four thousand years is made up of many links, each link being a repetition of the other. So long as a single pair of dogs resembling each other unite, so long will there be specimens of that species, simply because the children inherit the characteristics of the parents. So long as negroes marry with negroes, and Jews with Jews, so long must there be a perpetuation of the negro and Jewish types; but the tenth generation adds nothing to the evidence of the first, nor the ten thousandth to the tenth.

I believe that this fallacy, which destroys the whole value of the Cuvierian argument, has not before been pointed out; and even now you may perhaps ask if the fixity of species is not proved by the fact that like produces like? So far from this, that it is only by the aid of such a fact in organic nature that we can imagine *new* species to have arisen; in other words, those who believe in the variability of species, and the introduction of new forms by means of modification from the old, always invoke the law of hereditary transmission as the means of establishing accidental variations. Thus, let us suppose the Egyptian king to have had one hundred dogs, all of them staghounds, and no other form of dog to have existed at that time in that country; the dog species would be represented by the staghound. These staghounds would transmit to their offspring all their *specific characters*. But, as every one knows, however much dogs may resemble each other, they always present individual differences in size, color, strength, intelligence, etc. Now, if any one of these differences should happen to become marked, and to increase by the intermarriage of two dogs similarly distinguished by the marked peculiarity, this peculiarity would in time become established by hereditary transmission, and would form the starting-point of a new race of dogs—say the greyhound—unless it were obliterated by intermarriage with dogs of the old type. In the former case, we should have two races of dogs among the descendants of those figured on the Egyptian tombs; but as one of these races would still preserve the original staghound type, Cuvier would refer to *it* as a proof that species had not varied. We, on the other hand, should point to the greyhound as proof that animal forms *are* variable, and that a new form had arisen from modification of the old.

An objection will at once be raised to this illustration, to the effect that all zoologists admit the possibility of new varieties or races being formed; but they deny that new species can be formed. It is here that the equivoque of the word species prevents a clear understanding of each other's argument. Whiteness may justly be said to be unalterable; but white things may vary—they may become gray or yellow. In like manner species must be invariable, because species is a word indicating a particular group of characters; but animals may vary in these characters: they may present some of the characters less or more developed, and they may even want some of them. Now, as there is no absolute standard of what constitutes species, what sub-species, and what varieties, it becomes impossible to say whether any individual variation in an animal form shall constitute a new variety or a new species. With regard to dogs, the differences between the various races are so numerous and so marked as would suffice to constitute species, and even genera, in other groups of animals.

We must relinquish the idea of proving any thing by the paintings and sculptures of the ancients. When we find an Egyptian plow closely resembling the plow still in use in some places, we may identify it as of the same "species" as our own; but this does not disprove the fact that steam-plows, and plows of various construction, have been since invented, all of them being modifications of the original type. Formerly, and for many years, the stage-coach was our approved mode of conveyance—and it is still kept up in some districts; nevertheless, modifications of coachroad into tramroad, and tramroad into railroad, have gradually resulted in a mode of conveyance utterly unlike the stage-coach. It is the same with animals.

Let us never forget that species have no existence. Only individuals exist, and these *all vary* more or less from each other. When the modifications are slight, they have no name; when they are more marked, and are transmitted from one generation to another, they constitute particular races or varieties; when the differences are still more marked, they constitute sub-species; but, as Mr. Darwin says, "Certainly no clear line of demarkation has yet been drawn between species and sub-species; that is, the forms which in the opinion of some naturalists come very near to, but do not quite arrive at the rank of species; or again, between sub-species and well-marked varieties, or between lesser varieties and individual differences. These differences blend into each other in an insensible series; and a series impresses the mind with the idea of an actual passage." But the same process of divergence which establishes varieties out of individual differences, and species out of varieties, also serves to establish genera out of species, orders out of genera, and classes out of orders. It is doubtless difficult to conceive by what process of modification two animals of distinct genera, say a dog and a cat, were produced from a common stock; but organic analogies in abundance render it easy of belief. If we knew as much of zoology as we do of embryology, in respect of the affinities of divergent forms, it would be far less surprising that two different genera should arise from a common stock, than that all the various parts of the skeleton should arise from a common osseous element. We know that the jaws are identical with arms and legs—both being divergent modifications of a common osseous structure. We know that the arm of a man is identical with the fin of a whale or the wing of a bird. The differences here in form, size, and function are much greater than the differences which establish orders and classes in the animal series. Unless animal forms were modifications of some common type, it would be difficult to explain their remarkable affinities. As Mr. Darwin says, "It is a truly wonderful fact—the wonder of which we are apt to overlook from familiarity—that all animals and all plants throughout all time and space should be related to each other in group subordinate to group, in the manner which we every where behold, namely, varieties of the same species most closely related together, species of the same genus less closely and unequally related together, forming sections and sub-genera, species of distinct genera much less closely related, and genera related in different degrees, forming sub-families, families, orders, sub-classes, and classes. The several subordinate groups in any class can not be ranked in a single file, but seem rather to be clustered round points, and these round other points, and so on in almost endless circles. On the view that each species has been independently created, I can see no explanation of this great fact in the classification of all organic beings; but, to the best of my judgment, it is explained through inheritance, and the complex action of natural selection entailing extinction and divergence of character. The affinities of all the beings of the same class have sometimes been represented by a great tree. I believe this simile largely speaks the truth. The green and budding twigs may represent existing species, and those produced during each former year may represent the long succession of extinct species. At each period of growth all the growing twigs have tried to branch out on all sides, and to overtop and kill the surrounding twigs and branches, in the same manner as species and groups of species have tried to overmaster other species in the great struggle for life. The limbs divided into great branches, and these into lesser branches, were themselves once, when the tree was small, budding twigs; and this connection of the former and present buds by ramifying branches may well represent the classification of all extinct and living species in groups subordinate to groups. Of the many twigs which flourished when the tree was a mere bush, only two or three, now grown into great branches, yet survive and bear all the other branches; so with the species which lived during long-past geological periods, very few now have living and modified descendants. . . . As buds give rise by growth to fresh buds, and these, if vigorous, branch out and overtop on all sides many a feebler branch, so by generation, I believe, it has been with the great

Tree of Life, which fills with its dead and broken branches the crust of the earth, and covers the surface with its ever-branching and beautiful ramifications."*

It will not be expected that in these brief and desultory remarks I should touch on all, or nearly all, the important points in the discussion respecting the fixity of species. Mr. Darwin's book is in every body's hands, and my object has been to facilitate, if possible, the comprehension of his book, and the adoption of a more philosophical hypothesis, by pointing out the weakness of the chief argument on the other side. There is one more argument which may be noticed—the more so as it is constantly adduced with triumph by the one school, and admitted as a difficulty by the other. Its force is so great that it prevents many from accepting the development hypothesis. It is the argument founded on the alleged impossibility of hybrids continuing the race. More than two or three generations of hybrids, it is said, can never be maintained; after that, the new form perishes, thus clearly showing how Nature repudiates such amalgamations, and keeps her species jealously distinct and invariable. This argument is held to be the touchstone of the doctrine of species. I wish it were so; because, in that case, the question would no longer be one of hypothesis, since we have now the indubitable proof that some hybrids *are* fertile unto the thirteenth generation and onward.

* DARWIN: *Origin of Species*, p. 128.

A history of the various attempts which have been made to prove and disprove the fertility of hybrids would lead us beyond our limits; the curious reader is referred to the works cited below.* One decisive case alone shall be given here, and no one will dispute that it *is* decisive.

The hare (*lepus timidus*) is assuredly of a distinct species from the rabbit (*lepus cuniculus*). So distinct are these species, that any classification which should range them as one would violate every accepted principle. The hare is solitary, the rabbit gregarious; the hare lives on the surface of the earth, the rabbit burrows under the surface; the hare makes her home among the bushes, the rabbit makes a sort of nest for her young in her burrow— keeping them there till they are weaned; the hare has reddish-brown flesh, the rabbit white flesh; while the odor exhaled by each, and the flavor of each, are unmistakably different. The hare has many anatomical characters differing from those of the rabbit, such as greater length and strength of the hind legs, larger body, shorter intestine, thicker skin, firmer hair, and different color. The hare breeds only twice or thrice a year, and at each litter has only two or four; the rabbit will breed eight times a year, and each time has four, six, seven, and even eight young ones. Finally, the

---

* ISIDORE GEOFFROY ST. HILAIRE: *Hist. Nat. Générale des Règnes Organiques*, 1860, iii., 207 *sq*. BROCA: *Mémoire sur l'Hybridité*, in BROWN-SEQUARD'S *Journal de la Physiologie*, 1859.

two are violent foes: the rabbits always destroy the hares, and all sportsmen are aware that if the rabbits be suffered to multiply on an estate, there will be small chance of hares.

Nevertheless, between species so distinct as these, a new hybrid race has been reared by M. Rouy, of Angoulême, who each year sends to market upward of a thousand of his *Leporides*, as he calls them. His object was primarily commercial, not scientific. His experiments, extending from 1847 to the present time, have not only been of great commercial value—introducing a new and valuable breed—but have excited the attention of scientific men, who are now availing themselves of his skill and experience to help them in the solution of minor problems. It is enough to note here that these hybrids of the hare and the rabbit are fertile, not only with either hares or rabbits, but *with each other*. Thirteen generations have already been enumerated, and the last remains so vigorous that no cessation whatever is to be anticipated.

In presence of this case (and others, though less striking, might be named) there is but one alternative—either we must declare that rabbits and hares form one and the same species—which is absurd—or we must admit *that new types may be formed by the union of two existing races;* and, consequently, that species *are* variable. If the doctrine of fixity of species acknowledges the touchstone of hybridity, the fate of the doctrine is settled forever.

Although I conceive the doctrine of fixity of species to be altogether wrong, I can not say that the arguments adduced in favor of the development hypothesis rise higher than a high degree of probability, still very far from demonstration; they will leave even the most willing disciple beset with difficulties and doubts. When stated in general terms, that hypothesis has a fascinating symmetry and simplicity; but no sooner do we apply it to particular cases, than a thick veil of mystery descends, and our pathway becomes a mere blind groping toward the light. There is nothing but what is perfectly conceivable, and in harmony with all analogies, in the idea of all animal forms having arisen from successive modifications of one original form, but there are many things perfectly conceivable which have nevertheless no existence; there are many explanations perfectly probable which are not true; and when we come to seek for the evidence of the development hypothesis, that evidence fails us. It *may* be true, but we can not say that it *is* true. Ten years ago I espoused the hypothesis, and believed that it must be the truth; but ten years of study, instead of deepening, have loosened that conviction: they have strengthened my opposition to the hypothesis of fixity of species, but they have given greater force to the difficulties which beset the development hypothesis, and have made me feel that at present the requisite evidence is wanting. I conclude with reminding the reader that the question of the origin of species is at present incapable of a positive answer; of the two hypotheses, that of development seems the more harmonious with our knowledge; but it is no more than an hypothesis, and will probably forever remain one. Now an hypothesis, although indispensable as a provisional mode of grouping together facts, and giving them some sort of explanation, is, after all, only a *guess*, and it may be absurdly wide of the truth. In Natural History, as in all other departments of speculative ingenuity, there have been a goodly number of outrageously extravagant hypotheses gravely propounded and credulously accepted. Men prefer an absurd guess to a blank; they would rather have a false opinion than no opinion; and one of the last developments of philosophic culture is the power of *abstaining* from forming an opinion where the necessary data are absent.

If you wish to see how easily hypotheses are formed and accepted, you need only turn over the history of any science. If you want a laugh at credulity, read a chapter of Pliny's *Natural History*. Pliny is a classic, and was for centuries an authority; but, looked at with impartial eyes, he appears the veriest "old woman" that ever wrote in a beautiful style. He was a mere bookworm, without a particle of scientific insight. His was not an age when men had much regard to evidence; but to him the suspicion never seems to have occurred that Gossip Report could be given to romancing, or that travelers could "see strange things." No fable is too monstrous for his credulity.

One of the pretty fables Pliny repeats is that pearls are formed by drops of dew falling into the gaping valves of the oyster. It never occurred to him to ask whether oysters were ever exposed to the dew? whether the drops *could* fall into their valves? whether oysters kept their valves open except when under water? or, finally, whether, if the dew *did* fall in, it would *remain* a rounded drop? The drop of dew had a certain superficial resemblance to the pearl, and that was enough. Ælian's hypothesis was somewhat better: he supposed that the pearls were produced by lightning flashing into the open shells.

Turning from these ancient sages, you will ask how pearls are formed? And almost any ingenious modern, not a zoologist, will tell you (and tell you falsely) that the pearl is a disease of the oyster. One is somewhat fatigued at the merciless frequency with which this notion has been dragged in, as an illustration of genius issuing out of sorrow and adversity, and it is time to stop that "damnable iteration" by discrediting the notion. Know then that if

> "Most wretched men
> Are cradled into poetry by wrong,
> They learn in suffering what they teach in song"—

it is not true that oysters secrete in suffering what

women wear as necklaces. Disease would be the very worst cradle for pearls. The idea of disease originated in a fanciful supposition of pearls being to the oyster and mussel what gall-stones and urinary calculi are to higher and more suffering animals. Réaumur, to whom we owe so many good observations and suggestive ideas, came near the truth when, in 1717, he showed that the structure of pearls was identical with the structure of the shells in which they grow. He attributed their formation to the morbid effusion of coagulating shell-material.

I presume you know that shells are formed by a secretion from the *mantle?* The mantle is that delicate semi-transparent membrane which you observe, on opening a mussel, lining the whole interior of the shells, and having at its free margins a sort of fringe of delicate tentacles which are sensitive and retractile. A microscopic examination of these fringes shows them to be glandular in structure—that is, they are secreting organs. The whole mantle, indeed, is a secreting organ, and its secretion is the shell-material: the fringes secrete the coloring matters of the shell, and enlarge its *circumference;* the rest of the mantle secretes the nacre, or mother of pearl, and increases the *thickness* of the shell. Now it is obvious that the formation of pearl nacre and of pearls depends on the *healthy* condition of the mantle, not on its diseases. If the mantle be injured, the nacre is not secreted at all, or in less quantities.

But, although pearls depend upon the healthy, not the diseased activity of the mantle, it is clear that there must be some unusual condition present for their formation, since the secretion of nacre does not spontaneously assume the form of pearls. What is the unusual condition? Naturalists are at present divided into two camps, fighting vigorously for victory. The one side maintains that the origin of a pearl is this: an egg of the oyster has escaped and strayed under the mantle, or the egg of a parasite has been *deposited* there; this egg forms the nucleus round which the nacre forms, and thus we have the pearl. The other side maintains with great positiveness that *any thing* will form a nucleus, a grain of sand no less than the egg of a parasite. 'Tis a pretty quarrel, which we may leave them to settle. Some aver that grains of sand are more numerous than any thing else; but Möbius says that of forty-four sea pearls and fifteen fresh-water pearls examined by him, not one contained a grain of sand; and Filippi, who has extensively investigated this subject, denies that a grain of sand ever forms the nucleus of a true pearl. Both Filippi and Küchenmeister[*] declare that a parasite gets into the mussel or oyster, and its presence there stimulates an active secretion of nacre.

There are pearls, according to Möbius, which consist of three different systems of layers, like the shells in which they are formed; with this differ-

[*] See their interesting essays in MÜLLER's *Archiv.*, 1856.

ence, that these layers are *reversed;* in the shell the nacre forms the innermost layer, in the pearl it forms the outermost. Hence the qualities of the pearl depend on the shell, and on the different proportions of nacre and carbonate of lime.

Since we know how pearls are made, may it not be expected that we should learn to make them? Ever since the days of Linnæus the hope has been entertained, and it is now becoming every day more likely to be realized. Imperfect pearls have been made in abundance. The Chinese have long practiced the art. They simply remove the large freshwater mussel from the water, insert a foreign substance under the mantle, and in two or three years (if I remember rightly) they take the mussels up again, and find the pearls formed. In this way they make little mother-of-pearl Josses, which are sold for a penny each; and I remember seeing a couple of large shells in the Anatomical Museum at Munich, the whole length of which was occupied by rows of little squab Josses, very comical to behold. I was informed that a copper chain of these deities had been inserted under the mollusk's mantle, and this was the result.

CHAPTER VI.

Every Organism a Colony.—What is a Paradox?—An Organ is an independent Individual and a dependent one.—A Branch of Coral.—A Colony of Polypes.—The Siphonophora.—Universal Dependence.—Youthful Aspirings.—Our Interest in the Youth of great Men.—Genius and Labor.—Cuvier's College Life; his Appearance in Youth; his Arrival in Paris.—Cuvier and Geoffroy St. Hilaire.—Causes of Cuvier's Success.—One of his early Ambitions.—M. le Baron.—*Omnia vincit labor.*—Conclusion.

THAT an animal organism is made up of several distinct organs, and these the more numerous in proportion to the rank of the animal in the scale of beings, is one of those familiar facts which have their significance concealed from us by familiarity. But it is only necessary to express this fact in language slightly altered, and to say that an animal organism is made up of several distinct *individuals*, and our attention is at once arrested. Doubtless it has a paradoxical air to say so; but Natural History is full of paradoxes; and you are aware that a paradox is far from being necessarily an absurdity, as some inaccurate writers would lead us to suppose; the word meaning simply "contrary to what is thought"—a meaning by no means equivalent to "contrary to what is the fact." It is paradoxical to call an animal an aggregate of individuals, but it is so because our thoughts are not very precise on the subject of individuality—one of the many abstractions which remain extremely vague. To justify this application of the word individual to every distinct organ would be difficult in ordinary speech, but in philosophy there is ample warrant for it.

An organ, in the physiological sense, is an *instrument* whereby certain functions are performed. In the morphological sense, it arises in a *differentiation*, or setting apart, of a particular portion of the body for the performance of particular functions—a group of cells, instead of being an exact repetition of all the other cells, takes on a difference, and becomes distinguished from the rest as an organ.*

Combining these two meanings, we have the third or philosophical sense of the word, which indicates that every organ is an individual existence, dependent more or less upon other organs for its maintenance and activity, yet biologically distinct. I do not mean that the heart will live independent of the body—at least not for long, although it does continue to live and manifest its vital activity for some time after the animal's death; and, in the cold-blooded animals, even after removal from the body. Nor do I mean that the legs of an animal will manifest vivacity after amputation, although even the legs of a man are not dead for some time after amputation; and the parts of some of the lower animals are often vigorously independent.

* See on this point what was said in our first chapter, p. 22.

Thus I have had the long tentacles of a *Terebella* (a marine worm) living and wriggling for a whole week after amputation.* In speaking of the independence of an organ, I must be understood to mean a very dependent independence; because, strictly speaking, absolute independence is nowhere to be found; and, in the case of an organ, *it* is of course dependent on other organs for the securing, preparing, and distributing of its necessary nutriment. The tentacles of my *Terebella* could find no nutriment, and they perished from the want of it, as the *Terebella* itself would have perished under like circumstances. The frog's heart now beating on our table with such regular systole and diastole, as if it were pumping the blood through the living animal, gradually uses up all its force; and since this force is not replaced, the beatings gradually cease. A current of electricity will awaken its activity for a time, but at last every stimulus will fail to elicit a response. The heart will then be dead, and decomposition will begin.

Dependent, therefore, every organ must be on some other organs. Let us see how it is also independent; and for this purpose we glance, as usual, at the simpler forms of life to make the lesson easier. Here is a branch of coral, which you know to be in its living state a colony of polypes. Each of these multitudinous polypes is an individual, and each exactly resembles the others. But the whole

* *Seaside Studies*, 2d edition, p. 59 *sq.*

colony has one nutritive fluid in common. They are all actively engaged in securing food, and the labors of each enrich all. It is animal socialism of the purest kind—there are no rich and no poor, neither are there any idlers. Formerly the coral-branch was regarded as one animal—an individual; and a tree was and is commonly regarded as one plant—an individual. But no zoologist now is unaware of the fact that each polype on the branch is a distinct individual, in spite of its connections with the rest; and philosophic botanists are agreed that the tree is a colony of individual plants—not one plant.

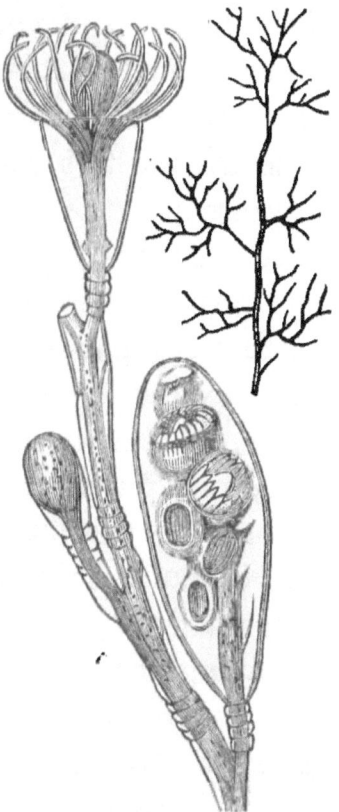

Fig. 20.—CAMPANULARIA (magnified and natural size).

Let us pass from the coral to the stem of some other polype, say a Campanularia. Above is the representation of such a stem, of the natural size, and beside it a tiny twig much magnified. You observe the ordinary polype issuing from one of the capsules, and expanding its coronal of tentacles in the water. The food it secures will pass along the digestive tract to each of the other capsules. Under the microscope you may watch this oscillation of the food. But your eye detects a noticeable difference between this polype in its capsule, and the six semi-transparent masses in the second capsule; although the two capsules are obviously identical, they are not the same; a *differentiation* has taken place. Perhaps you think that six polypes are here crowding into one capsule? Error! If you watch with patience, or if you are impatient yet tolerably dexterous, you may press these six masses out, and then will observe them swim away, so many tiny jellyfish. Not polypes at all, but jellyfish, are in this capsule; and these, in due time, will produce polypes, like that one now waving its tentacles.

Having made this observation, it will naturally occur to you that the polype stem which bore such different capsules as are represented by these two may perhaps be called a colony, but it is a colony of different individuals. While they have all one skeleton in common, nutrition in common, and respiration in common, they have at least one differentiation, or setting apart for a particular purpose, and that is the reproductive capsule. This is an individual as much as any of the others, but it is an individual that does nothing for the general good; it takes upon itself the care of the race, and becomes an "organ" for the community; the others feed it, and it is absolved from the labor of nutrition as much as the arm or the brain of a man are.

From this case, let us pass to the group of jellyfish called *Siphonophora* (siphon-bearers) by naturalists, and we shall see this union of very different individualities into one inseparable colony still more strikingly exhibited: there are distinct individuals to feed the colony, individuals to float it through the water, individuals to act as feelers, and to keep certain parts distended with fluid, and finally reproductive individuals. All these are identical in origin, and differ only by slight differentiations.* Here we have obviously an approach to the more complex organism in which various distinct organs perform the several functions, only no one calls the organism a colony.

The individuals composing one of these Siphonophora are so manifestly analogous to organs, that their individuality may perhaps be disputed, the more so as they do not live separately. But the gradations of separation are very fine. You would never hesitate to call a bee or an ant an individual, yet no bee or ant could exist if separated from its colony. So great is the "physiological division of labor" which has taken place among these insects, that one can not get food, another can not feed itself, but it will fight for the community; another can not work, but it will breed for the community; another can not breed, but it will work. Each of

* Compare LEUCKART: *Ueber den Polymorphismus der Individuen.* GEGENBAUR: *Grundzüge der Vergleichende Anatomie;* and HUXLEY's splendid monograph on the *Oceanic Hydrozoa*, published by the Ray Society.

these is little more than separated organs of the great insect organism, as the heart, stomach, and brain are *united* organs of the human organism. Remove one of these insects from the community, and it will soon perish, for its life is bound up with the whole.

And so it is every where; the dependence is universal:

> "Nothing in this world is single,
> All things, by a law divine,
> In one another's being mingle."

We are dependent on the air, the earth, the sunlight, the flowers, the plants, the animals, and all created things, directly or indirectly. Nor is the moral dependence less than the physical. We can not isolate ourselves if we would. The thoughts of others, the sympathies of others, the needs of others—these too make up our life; without these we should quickly perish.

It was a dream of the youth Cuvier that a History of Nature might be written which would systematically display this universal interdependence. I know few parts of biography so interesting as those which show us great men in their early aspirings, when dreams of achievements vaster than the world has seen fill their souls with energy to achieve the something they do afterward achieve. It is, unhappily, too often but the ambition of youth we have to contemplate; and yet the knowledge that after-life brought with it less of hope, less of devotion, and less of generous self-sacrifice, renders these early days doubly interesting. Let the abatement of high hopes come when it may, the existence of an aspiration is itself important. I have been lately reading over again the letters of Cuvier when an obscure youth, and they have given me quite a new feeling with regard to him.

There is a good reason why novels always end with the marriage of the hero and heroine; our interest is always more excited by the struggles than by the results of victory. So long as the lovers are unhappy or apart, and are eager to vanquish obstacles, our sympathy is active; but no sooner are they happy, than we begin to look elsewhere for other strugglers on whom to bestow our interest. It is the same with biography. We follow the hero through the early years of struggle with intense interest, and as long as he remains unsuccessful, baffled by rivals or neglected by the world, we stand by him and want him to succeed; but the day after he is recognized by the world our sympathy begins to slacken.

It is this which gives Cuvier's *Letters to Pfaff** their charm. I confess that M. le Baron Cuvier, administrator, politician, academician, professor, dictator, has always had but a very tepid interest for me, probably because his career early became a continuous success, and Europe heaped rewards upon

---

* *Lettres de Georges Cuvier à C. M. Pfaff*, 1788-92. Traduites de l'Allemand, par Louis Marchant, 1858.

him; whereas his unsuccessful rival, Geoffroy St. Hilaire, claims my sympathy to the close. If, however, M. le Baron is a somewhat dim figure in my biographical gallery, it is far otherwise with the youth Cuvier as seen in his letters; and as at this present moment there is nothing under our microscope which can seduce us from the pleasant volume, suppose we let our "studies" take a biographical direction?

"Genius," says Carlyle, "means transcendent capacity for taking trouble, first of all." There are many young gentlemen devoutly persuaded of their own genius, and yet candidly avowing their imperfect capacity for taking trouble, who will vehemently protest against this doctrine. Without discussing it here, let us say that, genius or no genius, *success* of any value is only to be purchased by immense labor: and in science, assuredly, no one will expect success without first paying this price. In Cuvier's history may be seen what "capacity for taking trouble" was required before his success could be achieved; and this gives these *Lettres à Pfaff* a moral as well as an interest.

It was in the Rittersaal of the Academia Carolina of Stuttgardt that Pfaff, the once famous supporter of Volta, and in 1787 the fellow-student of Cuvier, first became personally acquainted with him. Although they had been three years together at the same university, the classification of students there adopted had prevented any personal acquaintance.

Pupils were admitted at the age of nine, and commenced their studies with the classic languages. Thence they passed to the philosophical class, and from that they went to one of the four faculties—Law, Medicine, Administration, and Military Science. Each faculty, of course, was kept distinct; and as Pfaff was studying philosophy at the time Cuvier was occupied with the administrative sciences, they never met, the more so as the dormitories and hours of recreation were different. The academy was organized on military principles. The three hundred students were divided into six classes, two of which comprised the nobles, and the other four the bourgeoisie. Each of these classes had its own dormitory, and was placed under the charge of a captain, a lieutenant, and two inferior officers. These six classes, in which the students were entered according to their age, size, and time of admission, were kept separate in their recreations as in their studies. But those of the students who particularly distinguished themselves in the public examinations were raised to the rank of knights, and had a dormitory to themselves, besides dining at the same table with the young princes who were then studying at the university. Pfaff and Cuvier were raised to this dignity at the same time, and here commenced their friendship.

What a charm there is in school friendships, when youth is not less eager to communicate its plans and hopes than to believe in the plans and hopes of others; when studies are pursued in common, opinions frankly interchanged, and the superiority of a friend is gladly acknowledged, even becoming a source of pride, instead of being, as in after years, a thorn in the side of friendship! This charm was felt by Cuvier and Pfaff, and a small circle of fellow-students who particularly devoted themselves to Natural History. They formed themselves into a society, of which Cuvier drew up the statutes and became the president. They read memoirs, and discussed discoveries with all the gravity of elder societies, and even published, among themselves, a sort of *Comptes Rendus*. They made botanical, entomological, and geological excursions; and, still further to stimulate their zeal, Cuvier instituted an Order of Merit, painting himself the medallion: it represented a star, with the portrait of Linnæus in the centre, and between the rays various treasures of the animal and vegetable world. And do you think these boys were not proud when their president awarded them this medal for some happy observation of a new species, or some well-considered essay on a scientific question?

At this period Cuvier's outward appearance was as unlike M. le Baron as the grub is unlike the butterfly. Absorbed in his multifarious studies, he was careless about disguising the want of elegance in his aspect. His face was pale, very thin and long, covered with freckles, and encircled by a shock of red hair. His physiognomy was severe and melancholy. He never played at any of the boys' games, and seemed as insensible of all that was going on around him as a somnambulist. His eye seemed turned inward; his thoughts moved amid problems and abstractions. Nothing could exceed the insatiable ardor of his intellect. Besides his special administrative studies, he gave himself to Botany, Zoology, Philosophy, Mathematics, and the history of literature. No work was too voluminous or too heavy for him. He was reading all day long, and a great part of the night. "I remember well," says Pfaff, "how he used to sit by my bedside going regularly through Bayle's Dictionary. Falling asleep over my own book, I used to awake after an hour or two, and find him motionless as a statue, bent over Bayle." It was during these years that he laid the basis of that extensive erudition which distinguished his works in after life, and which is truly remarkable when we reflect that Cuvier was not in the least a bookworm, but was one of the most active *workers*, drawing his knowledge of details from direct inspection whenever it was possible, and not from the reports of others. It was here, also, that he preluded to his success as a professor, astonishing his friends and colleagues by the clearness of his exposition, which he rendered still more striking by his wonderful mastery with the pencil. One may safely say that there are few talents which are not available in Natural History; a talent for drawing is pre-eminently useful, since it not only enables a man to preserve observations of fugitive appearances, but sharpens his faculty of observation by the exercise it gives. Cuvier's facile pencil was always employed: if he had nothing to draw for his own memoirs or those of his colleagues, he amused himself with drawing insects as presents to the young ladies of his acquaintance—an entomologist's gallantry, which never became more sentimental.

In 1788, that is, in his nineteenth year, Cuvier quitted Stuttgardt, and became tutor in a nobleman's family in Normandy, where he remained till 1795, when he was discovered by the Abbé Tessier, who wrote to Parmentier, "I have just found a pearl in the dunghill of Normandy:" to Jussieu he wrote, "Remember it was I who gave Delambre to the academy; in another department this also will be a Delambre." Geoffroy St. Hilaire, already professor at the Jardin des Plantes, though younger than Cuvier, was shown some of Cuvier's manuscripts, which filled him with such enthusiasm that he wrote to him, "Come and fill the place of Linnæus here; come and be another legislator of natural history." Cuvier came, and Geoffroy stood aside to let his great rival be seen.

Goethe, as I have elsewhere remarked, has noticed the curious coincidence of the three great zoologists successively opening to their rivals the path to distinction: Buffon called Daubenton to aid him; Daubenton called Geoffroy; and Geoffroy called

Cuvier. Goethe further notices that there was the same radical opposition in the tendencies of Buffon and Daubenton as in those of Geoffroy and Cuvier—the opposition of the synthetical and the analytical mind. Yet this opposition did not prevent mutual esteem and lasting regard. Geoffroy and Cuvier were both young, and had in common ambition, love of science, and the freshness of unformed convictions. For, alas! it is unhappily too true, that just as the free communicativeness of youth gives place to the jealous reserve of manhood, and the youth who would only be too pleased to tell all his thoughts and all his discoveries to a companion would in after years let his dearest friend first see a discovery in an official publication, so likewise, in the early days of immature speculation, before convictions have crystallized enough to present their sharp angles of opposition, friends may discuss and interchange ideas without temper. Geoffroy and Cuvier knew no jealousy then. In after years it was otherwise.

Geoffroy had a position—he shared it with his friend; he had books and collections—they were open to his rival; he had a lodging in the museum—it was shared between them. Daubenton, older and more worldlywise, warned Geoffroy against this zeal in fostering a formidable rival, and one day placed before him a copy of Lafontaine open at the fable of *The Bitch and her Neighbor*. But Geoffroy was not to be daunted, and probably felt himself strong enough to hold his own. And so the two happy, active youths pursued their studies together, wrote memoirs conjointly, discussed, dissected, speculated together, and "never sat down to breakfast without having made a fresh discovery," as Cuvier said truly enough, for to them every step taken was a discovery.

Cuvier became almost immediately famous on his arrival at Paris, and his career henceforward was one uninterrupted success. Those who wish to gain some insight into the causes of this success should read the letters to Pfaff, which indicate the passionate patience of his studies during the years 1788–1795, passed in obscurity on the Norman coast. Every animal he can lay hands on is dissected with the greatest care, and drawings are made of every detail of interest. Every work that is published of any note in his way is read, analyzed, and commented on. Lavoisier's new system of chemistry finds in him an ardent disciple. Kielmeyer's lectures open new vistas to him. The marvels of marine life, in those days so little thought of, he studies with persevering minuteness and with admirable success. He dissects the cuttlefish, and makes his drawings of it with its own ink. He notes minute characters with the patience of a species-monger, whose sole ambition is to affix his name to some trifling variation of a common form, yet with this minuteness of detail he unites the largeness of view necessary to a comparative anatomist.

"Your reflections on the differences between animals and plants," he writes, "in the passage to which I previously referred, will be the more agreeable to me, because I am at present working out a new plan of a general natural history. I think we ought carefully to seek out the relation of all existences with the rest of nature, and, above all, to show their part in the economy of the great All. In this work I should desire that the investigator should start from the simplest things, such as air and water, and after having spoken of their influence on the whole, he should pass gradually to the compound minerals, from these to plants, and so on; and that at each stage he should ascertain the exact degree of composition, or, which is the same thing, the number of properties it presents over and above those of the preceding stage, the necessary effects of these properties, and their usefulness in creation. Such a work is yet to be executed. The two works of Aristotle, *De Historia Animalium*, and *De Partibus Animalium*, which I admire more each time that I read them, contain a part of what I desire, namely, the comparison of species, and many of the general results. It is, indeed, the first scientific essay at a natural history. For this reason it is necessarily incomplete, contains many inaccuracies, and is too far removed from a knowledge of physical laws." He passes on from Aristotle to Pliny, Theophrastus, Dioscorides, Aldovrandus, Gesner, Gaspar Bauhin, and Ray, rapidly sketching the history of natural history as a science, and concluding with this criticism on these attempts at a nomenclature which neglected real science: "These are the dictionaries of natural history; but when will the *language* be spoken?"

No one who reads these letters attentively will be surprised at the young Cuvier's taking eminent rank among the men of science in France; and Pfaff, on arriving in Paris six years afterward, found his old fellow-student had become "a personage." The change in Cuvier's appearance was very striking. He was then at his maturity, and might pass for a handsome man. His shock of red hair was now cut and trimmed in Parisian style; his countenance beamed with health and satisfaction; his expression was lively and engaging; and, although the slight tinge of melancholy which was natural to him had not wholly disappeared, yet the fire and vivacity of his genius overcame it. His dress was that of the fashion of the day, not without a little affectation. Yet his life was simple, and wholly devoted to science. He had a lodging in the Jardin des Plantes, and was waited on by an old housekeeper, like any other simple professor.

On Pfaff's subsequent visit things were changed. Instead of the old housekeeper, the door was opened by a lackey in grand livery. Instead of asking for "Citizen Cuvier," he inquired for Monsieur Cuvier; whereupon the lackey politely asked whether he wished to see M. le Baron Cuvier, or M. Fréderic, his brother? "I soon found where I was," continues Pfaff. "It was the baron, separated from me by that immense interval of thirty years, and by those high dignities which an empire offers to the ambition of men." He found the baron almost exclusively interested in politics, and scarcely giving a thought to science. The "preparations" and "injections" which Pfaff had brought with him from Germany as a present to Cuvier were scarcely looked at, and were set aside with an indifferent "that's good," and "very fine;" much to Pfaff's distress, who doubtless thought the fate of the Martignac ministry an extremely small subject of interest compared with these injections of the lymphatics.

But it is not my purpose to paint Cuvier in his later years. It is to the studies of his youth that I would call your attention, to read there, once again, the important lesson that nothing of any solid value can be achieved without entire devotion. Nothing is earned without sweat of the brow. Even the artist must labor intensely. What is called "inspiration" will create no works, but only irradiate works with felicitous flashes; and even inspiration mostly comes in moments of exaltation produced by intense work of the mind. In science, incessant

G

and enlightened labor is necessary, even to the smallest success. Labor is not all; but without it genius is nothing.

With this homily, dear reader, may be closed our First Series of Studies; to be resumed hereafter, let me hope, with as much willingness on your part as desire to interest you on mine.

THE END.

A LIST OF BOOKS.

PUBLISHED BY

HARPER & BROTHERS, Franklin Square, N.Y.

LORD ELGIN'S MISSION TO CHINA, &c.

Narrative of Lord Elgin's Mission to China and Japan in 1857, '58, '59. Ly LAURENCE OLIPHANT, Secretary to Lord Elgin. Illustrations. 8vo, Muslin, $2 75.

OLD LEAVES.

Old Leaves: Gathered from Household Words. By W. HENRY WILLS. 12mo, Muslin, $1 00.

HORACE.

Horatius, ex recensione A. J. MACLEANE, A.M. 18mo, Flexible Binding, 40 cents.

ÆSCHYLUS.

Æschylus, ex novissima recensione F. A. PALEY, A.M. 18mo, Flexible Binding, 40 cents.

THE CAXTONS.

A Family Picture. By Sir EDWARD BULWER LYTTON, Bart. Library Edition. 12mo, Muslin.

SQUIER'S NICARAGUA.

Nicaragua: Its People, Scenery, Monuments, Resources, Conditions, and proposed Canal. With One Hundred Maps and Illustrations. By E. G. SQUIER, formerly Chargé d'Affaires of the United States to the Republics of Central America. A Revised Edition. Maps and Illustrations. 8vo, Muslin. (Just Ready.)

HARPER'S SCHOOL AND FAMILY READERS.

Harper's Series of School and Family Readers. By MARCIUS WILLSON.

☞ The Primer and first Four Readers are now passing through the Press. All are splendidly Illustrated. The Prices will be announced shortly.

LUCY CROFTON.

By the Author of "Margaret Maitland," "The Laird of Norlaw," "The Days of My Life," &c., &c. 12mo, Muslin, 75 cents.

LIFE AND TIMES OF GEN. SAM. DALE,

The Mississippi Partisan. By J. F. H. CLAIBORNE. Illustrated by JOHN McLENAN. 12mo, Muslin, $1 00.

LIFE IN SPAIN:

Past and Present. By WALTER THORNBURY. With Illustrations. 12mo, Muslin, $1 00.

SELF-HELP;

With Illustrations of Character and Conduct. By SAMUEL SMILES, Author of "The Life of George Stephenson." 12mo, Muslin, 75 cents.

PREACHERS AND PREACHING.

By KIRWAN, Author of "Letters to Bishop Hughes," "Romanism at Home," "Men and Things in Europe," &c., &c. 12mo, Muslin, 75 cents.

THE VIRGINIANS.

A Tale of the Last Century. By W. M. THACKERAY, Author of "The Newcomes," "Vanity Fair," "Pendennis," &c., &c. With Illustrations by the Author. 8vo, Paper, $1 95; Muslin, $2 00.

www.ingramcontent.com/pod-product-compliance
Lightning Source LLC
Chambersburg PA
CBHW020053170426
43199CB00009B/272